Instructor's Resource Guide

for

UNDERSTANDABLE STATISTICS

Fifth Edition
Brase/Brase

Charles Henry Brase
Regis University

Corrinne Pellillo Brase
Arapahoe Community College

D.C. Heath and Company
Lexington, Massachusetts   Toronto

Please address editorial correspondence to:

D.C. Heath and Company
Mathematics Editorial
125 Spring Street
Lexington, MA 02173

Published simultaneously in Canada.

Printed in the United States of America.

International Standard Book Number 0-669-39478-5

10 9 8 7 6 5 4 3 2 1

Preface

This instructor's resource guide is written to accompany the text *Understandable Statistics: Concepts and Methods*, fifth edition.

Part I contains suggestions for using the text and some alternate ways to order the topics and chapters on the basis of knowledge required for prerequisite chapters or sections are provided in the first chapter of this guide.   Also included is a chart correlating the text to the *Against All Odds* video series

Part II offers new chapter tests (two per chapter). The chapter tests are arranged so that they may be photocopied easily for classroom use. Each chapter test is keyed to a specific chapter, and answers with key steps are included at the end of Part II.   Faculty members wishing to test over several chapters may of course choose problems from chapter tests and combine them in a suitable way. These test questions also are included with many other test questions in the Test Item File supplement and ESATEST III Computerized Testing software that accompany the text.

Part III contains answers to all even-numbered problems in the text, and some of the key steps necessary to arrive at the answer. These answers also are included in the Instructor's Annotated Edition.   Answers to odd-numbered problems do not appear in this guide because they are provided in Appendix III of the text. The Study and Solutions Guide for students contains complete solutions to nearly half of the odd-numbered problems.

Part IV contains transparency masters of some key diagrams from the text. In addition are some screen displays from the Texas Instruments TI-82 graphics calculator some printouts from the computer software package *Computer Stat* which was written to coordinate with the text *Understandable Statistics* and some screen printouts from the professional statistical software package MINITAB.

Part V contains masters for the formula card as well as all of the tables included in Appendix II of *Understandable Statistics*, fifth edition. Using these masters, a faculty member can create custom handouts for student use during tests.

Supplements specifically designed to accompany the fifth edition of *Understandable Statistics* by Brase and Brase are

1. An *Annotated Instructor's Edition* containing margin comments and answers to even numbered problems. The answers to odd numbered problems are included in Appendix III of the Instructor's Edition as well as in the standard edition of the text used by students.

2. *Technology Guide* featuring

    Texas Instruments TI-82
    ComputerStat
    MINITAB

    This guide gives instructions for using the listed technologies as well as lab activities coordinated to the text *Understandable Statistics*.

3. ComputerStat This is a computer software package available for both MS- DOS and Macintosh. This software was written by the authors Brase and Brase to coordinate with the text *Understandable Statistics*. It offers options to use built in classroom demonstrations or to enter your own data.   Institutions adopting the text qualify for complimentary site licenses for this software. Contact D.C. Heath for details.

4. DATA DISK for MS-DOS and Macintosh This diskette contains real world data in MINITAB portable worksheets.   The data on the disk is described in the *Using Technology Guide*. Some of the lab activities for MINITAB in the *Using Technology Guide* incorporate the data on the DATA DISK.

5. *Study and Solutions Guide* by Elizabeth Farber of Bucks County Community College helps students reinforce their skills. Step by step solutions to about half of the odd-numbered problems are included. In addition Thinking About Statistics sections help students put ideas statistical ideas in context.

6. D.C. Heath ESATEST III Computerized Testing for MS-DOS and Macintosh contains a computerized test bank and test design platform. Faculty can custom design their own exams utilizing test file items and their own test questions as well.

7. *Test Item File* contains a printed copy of the same questions as those included in the D.C. Heath ESATEST III for *Understandable Statistics*, fifth edition..

Contents

Part I

Suggestions for Using the Text *Understandable Statistics,*
 fifth edition
Alternate Paths Through the Text
Correlation Chart for the *Against All Odds* Video Series

Suggestions for Using the Text

In writing this text we have followed the premise that a good textbook must be more than just a repository of knowledge. A good textbook should be an agent interacting with the student to create a working knowledge of the subject. To help achieve this interaction we have modified the traditional format in a way designed to encourage active student participation.

Each chapter opens with Chapter Overviews that provide non technical capsule summaries of each section to be presented in the chapter. Next is a Focus problem utilizing real world data. The Focus problem shows the student the kinds of questions they can answer when they have mastered the material in the chapter. In fact, students are asked to solve the Chapter Focus Problem as soon concepts required for the solution have been introduced.

A special feature of this text is the occurrence of built in Guided Exercises within the reading material of the text. These Guided Exercises with their completely worked out solutions help the students focus on key concepts of newly introduced material. The Section Problems reinforce student understanding and sometimes require the student to look at the concepts from a slightly different perspective than that presented in the section. Chapter problems are much more comprehensive. They require that students place the problem in the context of all that they have learned in the chapter. Data Highlights at the end of each chapter ask students to look at data as presented in newspapers, magazines, and other media and then to apply releavant methods of interpretation. Finally, Linking Concept problems ask students to verbalize their skills and synthesize the material.

We believe that the approach from small step Guided Exercises to Section Problems to Chapter Problems to to Data Highlights to Linking Concepts will enable the professor to employ his or her class time in a very profitable way, going from specific mastery details to more comprehensive decision-making analysis.

Calculators and statistical computer software relieve much of the computation burden from statistics. Many basic scientific calculators provide the mean and standard deviaiton. Those supporting two vairable statistics provide the coefficients for the least squares line, value of the corerplation coefficientient, and the predicted vlaue of y for a given x. Graphing calculators sort the data and many provide the median and quartile values. Some produce histograms, box plots, scatter plots and plot the least squares line. Statistical software packages give full support for descriptive statistics and inferential statistics. Students benefit from using these techonoligies. In many examples and exercises in *Understandable Statistics* we ask students to use calculators to verify answers. In text displays show TI-82 graphics calculator screens, Minitab printouts and ComputerStat printouts so that students may see the types of information readily available to them through use of technology.

However, it is not enough to enter data and punch a few buttons to get statistical results. The formulas producing the statistics contain a gread deal of information about the *meaning* of the statistics. The text breaks down formulas into tabular format so that students can see the information in the formula. We find that it is useful to take class time to discuss formulas. For instance, an essential part of the standard deviation formula is the comparison of each data value to the mean. When we point this out to students, it gives meaning to the standard deviation. When students understanda the content of the formulas, the numbers they get from their calculator or computer begin to make sense and have meaning.

For courses in which technologies are more incorporated into the curriculum, we provide a separate supplement, the *Technology Guide*. This guide gives specific hints for using the technologies, and specific lab activities to help students explore various statistical concepts.

Alternate Paths Through the Text

As previous editions, the fifth edition of *Understandable Statistics* is designed to be flexible. In most one semester courses, it is not possible to present all of the topics. However, the text provides many topics so that you can tailor a course to fit your students' needs. The text provides students a *readable reference* for topics not specifically included in your course.

For courses needing more descriptive statistics do all of Chapters 1, 2, and 3. These chapters give a good overview of sampling, gathering data, data types, and graphs including circle graphs, bar graphs, Pareto charts, time plots, histograms, relative frequency histograms, frequency polygons, ogives, stem-and-leaf plots, and box-and-whisker plots. Skewness and symmetry of histograms is also discussed. Summary statistics includethe mean, median, mode trimmed mean, sample standard deviation, population standard deviation, range, coefficient of variation, Chebyshev's Theorem, quartiles and, percentiles. Also included are the mean and standard deviation for grouped data and weighted averages.

For courses with <u>limited time for descriptive statistics,</u> have students read Chapter 1, and include Section 2.1 on random samples, part of Section 2.3 including histograms and relative frequency histograms, Section 3.1 on averages (mean, median, mode), part of Section 3.2 including the range and standard deviation. Other topics such as stem-and-leaf plots in Section 2.4 and box-and-whisker plots in Section 3.4 may be included as well, depending on the course objectives.

Probability can be done on two levels. For a <u>brief introduction</u>, do Sections 4.1, part of 4.2 (not including Further Examples), and Section 4.4. In Section 4.2 assign only the very routine problems emphasizing mutually exclusive events and independent events. For a more in depth coverage of probability do all of Chapter 4, including the counting techniques of Section 4.3 Spend more time in Section 4.2 which contains a wealth of problems using the addition and multiplication rules, and conditional probability. Also consider introducing Bayes' Theorem from Appendix I.

For courses needing to spend more time on inferential statistics, go light in Chapters 2, 3, and 4. Cover only Sections 2.1, 2.3, 3.1, 3.2, 4.1, part of 4.2, and 4.4. Then spend your time in Chapters 5 through 11.

For courses needing <u>linear regression early</u>, begin Chapter 10 after Chapter 3. Do Sections 10.1 which includes scatter plots, part of Section 10.2 which includes the least squares line and the standard error of estimate (stopping before confidence intervals for prediction), Section 10.3 which includes the correlation coefficient. Later, after discussion of confidence intervals in Chapter 8 and the concept of hypothesis testing, finish Section 10.2 and present Section 10.4. The instructions given for the section

Understandable Statistics, 5th edition
Linked to
Against All Odds: Inside Statistics

The television series *Against All Odds: Inside Statistics* produced by Annenberg/CBS consists of 26 programs, each 30 minutes in length. The video programs provide interesting applications of statistics on actual location. Some of the programs expand the discussion of topics found in the text *Understandable Statistics*.

Against All Odds videos correlate to the text *Understandable Statistics*, 5th edition in the following way.

Chapter *Understandable Statistics*	Program *Against All Odds*
Chapter 1 Getting Started	Program 1 What is Statistics?
Chapter 2 Organizing Data 2.1 Random Samples	Program 12 Experimental Design Program 13 Experiments and Samples
2.3 Histograms and Frequency Distributions 2.4 Stem-and-Leaf Displays	Program 2 Picturing Distributions

Chapter 3   Averages and Variation
  3.1   Mode, Median, and Mean
  3.2   Measures of Variation
	Program 3 Numerical Description of Distributions

Chapter 4   Elementary Probability Theorem
  4.1   What is Probability
  4.2   Some Probability Rules
	Program 15 What is Probability?

4.4 Introduction to Random Variables and Probability	Program 16 Random Variables

Chapter 5   The Binomial Probability Distribution and Related Topics
  5.1   Binomial Experiments
  5.2   The Binomial Distribution
  5.3   Additional Properties of the Binomial Distribution
	Program 17 Binomial Distributions

Chapter *Understandable Statistics*	Program *Against All Odds*
Chapter 6 Normal Distributions	
6.1 Graphs of Normal Probability Distributions	Program 4 Normal Distributions
6.2 Standard Units and The Standard Normal Distribution	
6.3 Areas Under the Standard Normal Distribution	Program 5 Normal Calculations Note: Table referred to gives areas in the tails of the distribution rather than from 0 to z.
6.4 Areas Under Any Normal Curve	
Chapter 7 Introduction to Sampling Distribution	
7.1 Sampling Distributions	Program 14 Samples and Sampling Distributions
7.2 Central Limit Theorem	Program 18 The Sample Mean and Control Charts Note: These are $\bar{x}$ control charts rather than x control charts as discussed in Section 6.1 of the text.
Chapter 8 Estimation	
8.1 Estimating μ with Large Samples	Program 19 Confidence Intervals
8.4 Choosing the Sample Size	
Chapter 9 Hypothesis Testing	
9.1 Introduction to Hypothesis Testing	Program 20 Significance Tests
9.2 Tests Involving the Mean μ Large Samples	Program 21 Inference for One Mean
9.3 The P Value in Hypothesis Testing	
9.4 Tests Involving the mean μ (Small Samples)	
9.5 Tests Involving A proportion	Program 22 Comparing Two Means
9.6 Tests Involving Paired Differences (Dependent Samples)	Program 23 Inference for Proportions
9.7 Testing Differences of Two Means or Two Proportions	

Part II
Chapter Tests
Answers to Chapter Tests

==

Chapter Test 1A

1. To estimate the number of hours of sleep for college students enrolled for 15 credits sleep on Tuesday nights, a sample of 50 students were surveyed.
   a) What is the implied population?
   b) What is the sample?
   c) Would this sample necessarily reflect the number of hours of sleep college students get on Friday night? Explain your answer.

2. A sales employee evaluation report contains the following information
   a) Dollar amount of sales for the past six months
   b) Dates of sales orders exceeding $5000
   c) Sales district
   d) Performance rating on a scale showing unsatisfactory, satisfactory and excellent
   e) Knowledge about product from a scale 1, 2, 3, 4, 5 with 5 being highest
   For the information in parts (a) through (e), list the highest level of measurement as ratio, inferential, ordinal, or nominal and explain your choice.

3. Categorize the style of gathering data (sampling, experiment, simulation, census) described in each of the following situations:
   a) Look at all the apartments in a complex and determine the monthly rent charged for each unit.
   b) Given one group of students a flu vaccination and compare the number of times these students are sick during the semester with students in a group who did not receive the vaccination.
   c) Select a sample of students and determine the percentage who are taking mathematics this semester.
   d) Use a computer program to show the effects on traffic flow when the timing of stop lights is changed.

4. Write a brief essay in which you describe what is meant by an experiment. Give an example of a situation in which data is gathered by means of an experiment. How is gathering data from an experiment different from using a sample from a specified population?

Chapter Test 1B

1. The county sheriff wants to know the proportion of drivers who make an illegal left turn as they leave the post office parking lot. A sample of 200 cars are observed.
    a) What is the implied population?
    b) What is the sample?
    c) Would this sample necessarily reflect the proportion of cars making illegal left turns at all intersections? Explain your answer.

2. A restaurant manager is developing a clientele profile. Some of the information for the profile follows:
    a) Gender of diners.
    b) Size of groups dining together
    c) Time of day the last diner of the evening departs
    d) Age grouping: young, middle age, senior
    e) Length of time a diner waits for a table

    For the information in parts (a) though (e), list the highest level of measurement as ratio, interval, ordinal, or nominal and explain your choice.

3. Categorize the style of gathering data (sampling, experiment, simulation, census) for the following situations:
    a) Consider all the students enrolled at your college this semester and report the age of each student.
    b) Select a sample of new F10 pickup trucks and count the number of manufacturer defects in each of the trucks.
    c) Use computer graphics to determine the flight path of a golf ball when the position of the hand on the golf club is changed.
    d) Teach one section of English composition using a specific word processing package and teach another without using any computerized word processing. Count the number of grammar errors made by students in each section on a final draft of a 20 page term paper.

4. Write a brief essay in which you discuss some of the aspects of surveys. Give specific examples to illustrate your main points.

Chapter Test 2A

1. Describe how you could use the random number table to simulate the outcomes (total number of dots showing on the top faces) of rolling two dice 4 times. Using the following row of random numbers from the table, find the first outcome.

    28703   51709   94456   48396   73780

2. A survey of students using a new automated telephone registration process identified the following complaints about the system. There were 350 complaints about the registration phone line being busy. Lack of advising information produced 100 complaints, difficulty in entering course selection code correctly produced 35 complaints. Difficulty in changing a previous course selection generated 120 complaints. Make a Pareto chart of this information.

3. In a health care utilization study, a random sample of 25 patients having total care insurance at a designated clinic was studied. The number of times each patient went to the clinic during a one year period was recorded. The results follow:

15	7	3	13	1	22	12	4	3	8
12	15	11	19	3	2	5	3	4	7
2	13	6	5	4					

    a) Make a frequency table with five classes showing class boundaries, frequencies, relative frequencies, and cumulative frequencies.
    b) Make a frequency histogram with five classes.
    c) Make a relative frequency histogram with five classes.
    d) Make an ogive with five classes.

4. The Reeling Angels, a popular recording group, gave a single concert in Littletown, Maine. The concert tickets sold out in the first few hours of availability. Scalper ticket agents purchased blocks of the best Section A tickets and resold them at much higher prices. A random sample of 30 people buying tickets from different scalpers showed the following prices paid per ticket (in dollars):

20	35	40	25	31	16	37	50	35	35
25	28	24	38	40	45	30	35	33	37
25	36	39	45	30	35	33	27	35	40.

    Make a stem-and-leaf display of the data.

5. Identify each of the following samples by naming the sampling technique used (cluster, convenience, random, stratified, systematic).
    a) Measure the length of time every fifth person coming into a bank waits for teller service over a period of two days.
    b) Take a sample of five zip codes from the Chicago metropolitan region and

select a sample of elementary schools from each of the zip code regions. Determine the number of students enrolled in first grade in each of the schools selected.

c) Divide the users of the computer online service Internet into different age groups and then select a random sample from each age group to survey about the amount of time they are connected to Internet each month.

d) Survey five friends regarding their opinion of the student cafeteria.

e) Pick a random sample of students enrolled at your college and determine the number of credit hours they have each accumulated toward their degree program.

6. Last year, the daily high temperatures for each day (beginning on Sunday) of the first week in July in Denver were:

      83   92   95   89   91   95   97

Make a time plot for these data.

Chapter Test 2B

1. A business employs 736 people. Describe how you could get a simple random sample of size 30 to survey regarding desire for professional training opportunities. Identify the first 5 to be included in the sample using the following random number sequence

   71886   56450   36507   09395   96951

2. A customer satisfaction survey identified the following complaints about Fair's Auto Repair. There were six complaints about the quality of work, 30 about the difficulty in getting an appointment, 40 about the timeliness of work completion and 2 about the courtesy of personnel. Make a Pareto chart of this information.

3. A random sample of 25 orders at fast food restaurants were studies to determine the dollar amount (rounded to the nearest dollar) of the order. Note that some of the orders included meals for several people. The results were

12	33	18	13	15	2	23	17	9	11
5	7	12	16	25	29	4	23	7	3
27	14	14	9	6					

   a) Make a frequency table with six classes showing class boundaries, frequencies, relative frequencies, and cumulative frequencies.
   b) Make a histogram with six classes.
   c) Make an ogive with six classes.
   d) Describe how a relative frequency histogram will be similar to a histogram and how it will be different.

4. Dream Travel Tours contracted with Eco Rent-a-Car to provide rental cars to travelers. Eco went out of business and did not provide the prepaid rental cars to 20 travelers. The travelers demanded refunds from Dream Travel. The times elapsed between refund request and the actual refunds were as follows (in days):

26	49	22	58	43	37	48	32	21	54
42	63	27	33	47	45	67	21	48	57

   Make a stem-and-leaf display of the data.

5. To determine monthly rental prices of apartment units in the San Francisco area, samples were constructed in the following ways. Categorize (cluster, convenience, simple random, stratified, systematic) each sampling technique described.
   a) Number all the units in the area and use a random number table to select the apartments to include in the sample.
   b) Divide the apartment units according to number of bedrooms and then sample from each of the groups.

    c) Divide the apartment units according to zip code and then sample from each of the zip code regions.

    d) Look in the newspaper and consider the first sample of apartment units that list rent per month.

    e) Call every 50th apartment complex listed in the yellow pages and record the rent of the unit with unit number closest to 200.

6. A video rental outlet has kept track of the volume of video rentals over the past week, beginning with Monday. The number of rentals for each day last week beginning with Monday follow. Make a time plot for these data.

    275   300   180   250   400   525   250

Chapter Test 3A

1. Response times for eight emergency police calls in Denver were measured to the nearest minute and found to be:

   7   10   8   5   8   6   8   9

   Find the mean, median, and mode.

2. At the University of Colorado a random sample of five faculty gave the following information about the number of hours spent on committee work each week:

   3   6   4   1   5

   a) Find the range.
   b) Find the sample mean.
   c) Find the sample standard deviation

3. Home Security Systems is studying the time utilization of its sales force. A random sample of 40 sales calls showed that the representatives spend an average of $\bar{x} = 42$ minutes on the road with standard deviation $s = 5$ minutes for each sales call.
   a) Compute the CV for this data.
   b) Use Chebyshev's Theorem to find the smallest time range in which we can expect *at least* 75% of the data to fall.

4. A psychology test to measure memory skills was given to a random sample of 43 students. The results follow, where x is the student score and f is the frequency with which students obtained this score.

x	0 - 10	11 - 21	22 - 32	33 - 43	44 - 54
f	1	12	18	9	3

Use the midoints to estimate (a) the mean and (b) the sample standard deviation of scores.

5. Advantage VCR's carry a one-year warranty. However, the terms of the warranty require that a registration card be sent in. One of the questions on the registration card asks the owner's age. For a random sample of 30 registration cards, the ages given were:

38	27	41	31	31	35	21	36	37	32
45	47	32	26	23	19	27	38	47	51
52	55	46	33	27	22	25	27	31	34

 a) Give the five number summary including the low value, Q_1, median, Q_3, high value.

 b) Make a box-and-whisker plot of the ages.

6. In Economics 210, weights are assigned to required activities as follow:

 project, 30%   test 1, 15%   test 2, 15%   final exam, 40%

 Each activity is graded on a 100 point scale. Bill earned 70 points on the project, 90 on test 1, 83 on test 2, and 92 on the final exam. Compute his overall weighted average in the economics class.

Chapter Test 3B

1. A package delivery service provided the following information about the weights of 20 packages chosen at random (weights to the nearest ounce):

50	55	18	21	64	32	21	52	33	41
21	60	18	21	37	40	8	16	21	18

   Find the mean, median, and mode of the weights.

2. A random sample of six credit card accounts gave the following information about the payment due on each card:

   $53.18 $71.12 $115.10 $27.30 $36.19 $66.48

   a) Find the range.
   b) Find the sample mean.
   c) Find the sample standard deviation.

3. Job Finders did a study of midmanagement jobs in large cities. A sample of such jobs showed the average number of applicants for each position to be $\bar{x} = 37$ with standard deviation $s = 6$.
   a) Compute the CV for the number of applicants for each position.
   b) Use Chebyshev's theorem to find the smallest time range in which we can expect *at least* 75% of the data to fall.

4. Market surveys are sometimes used to set the market price of a new consumer item. A random sample of 335 adults in Chicago was shown a new type of home video-telephone not yet on the consumer market. After inspecting and testing the new video-telephone, each person was asked how much he or she would be willing to pay for this video-telephone. The frequency results follow, where x is the cost and f is the number of people who said that the given cost was the highest amount they would pay for the item.

x	$50	$150	$250	$350	$450	#550
f	12	67	115	95	40	6

   Find (a) the mean and (b) the sample standard deviation for the amount these people are willing to pay for a home video-telephone.

5. McElroy Discount Fashions claim to have a fairly low mark-up percent on the items they sell. A random sample of 20 items showed the mark-up percent over cost to be (in percents):

28	35	47	42	51	15	72	33	27	22
37	36	51	29	36	72	51	55	68	53

   a) Give the five number summary including the low value, Q_1, median, Q_3, high value.
   b) Make a box-and-whisker plot of the ages.

6. Home Storage Units makes metal book shelves, cabinets, tool boxes, and so on. A satisfaction rating is given to products using the following weights

   Quality of final product, 5   Instructions for assembly, 3
   Completeness of package, 5

   Each quality is measured on a rating scale of 5 points, with 5 being the highest rating. One item received a rating of 4 for quality, 2 for instructions, and 5 for completeness of package. Compute a weighted average to determine the overall satisfaction rating.

Chapter Test 4A

1. The Aim n' Shoot Camera Company wants to estimate the probability that one of their new cameras is defective. A random sample of 400 new cameras shows 24 are defective.
   a) How would you estimate the probability that a new Aim n' Shoot camera is defective? What is your estimate?
   b) What is your estimate for the probability that an Aim n' Shoot camera is *not* defective?
   c) Either a camera is defective or it is not. What is the sample space in this problem? Do the probabilities assigned to the sample space add up to 1?

2. An urn contains 8 balls identical in every way except in color. There are 2 red balls, 5 green balls, and 1 blue ball.
   a) You draw 2 balls from the urn but replace the first before drawing the second. Find the probability that the first ball is red *and* the second is green.
   b) Repeat part a, but do not replace the first ball before drawing the second.

3. June wants to become a policewoman. She must take a physical exam and then the written one. Records of past cadets indicate that the probability of passing the physical exam is 0.82. Then the probability that a cadet passes the written exam *given* he or she has passed the physical exam is 0.58. What is the probability that June passes both exams?

4. A real estate surveyed a random sample of 70 clients. One question asks if the client is satisfied or not and another asks if the client was referred to the company by another client. Results of the survey are shown in the following table.

	Referred	Not Referred	Total
Satisfied	25	25	50
Not Satisfied	5	15	20
Total	30	40	70

   If a client is selected at random (from this group of 70 clients), find the probability
   a) The client is satisfied.
   b) The client is satisfied *and* was referred
   c) The client is satisfied *given* the client was referred
   d) Are the events satisfied and referred independent or not? Explain.

5. One make of cellular telephone comes is 3 models. Each model comes in 2 colors (dark green and white). If a store wants to display each model in each color, how many cellular telephones must be displayed? Make a tree diagram showing the outcomes for selecting a model and a color.

6. A local taxi company is interested in the number of pieces of luggage a cab caries on a taxi run. A random sample of 260 taxi runs gave the following information, where x = number of pieces of luggage and f = frequency with which taxi runs carried this many pieces of luggage.

x	0	1	2	3	4	5	6	7	8	9	10
f	42	51	63	38	19	16	12	10	6	2	1

    a) If a taxi run is chosen at random from these 260 runs, use relative frequencies to find P(x) for x = 0, 1, 2, 3, 4, 5, 6, 7, 8, 9, 10.

    b) Use a histogram to graph the probability distribution of part (a).

    c) Assuming these 260 taxi runs represent the population of all taxi runs in this area, what do you estimate the probability is that a randomly selected run will have from 0 to 4 pieces of luggage (including 0 and 4)?

    d) What do you estimate the probability is that a randomly selected taxi fun will have from 6 to 10 pieces of luggage (including 6 and 10)?

    e) Compute the expected value of the x distribution.

    f) Compute the standard deviation of the x distribution.

Chapter Test 4B

1. The city council has three liberal members (one of whom is the mayor) and two conservative members. One member is selected at random to testify in Washington, D.C.
    a) What is the probability the member is liberal?
    b) What is the probability the member is conservative?
    c) What is the probability the mayor is chosen?

2. An urn contains 9 balls identical in every way except in color. There are 3 red balls, 4 blue balls, and 2 white balls.
    a) You draw 2 balls from the urn but replace the first before drawing the second. Find the probability that the first ball is white *and* the second one is blue.
    b) Repeat part a but do not replace the first ball before drawing the second.

3. The dean of women at Brookfield College found that 20% of the female students are majoring in engineering. If 53% of the students at Brookfield are women, what is the probability that a student chosen at random will be a woman engineering major?

4. Student Life did a survey of students in which they asked if the student is part time or full time. Another question asked if the student voted or not in the most recent student elections. The results follow.

	Part Time Student	Full Time Student	Total
Voted	15	20	35
Did not vote	25	30	55
Total	40	50	90

If a student is selected at random (from this group of 90 students), find the probability
    a) The student voted in the most recent election.
    b) The student voted in the most recent election *and* is a part time student
    c) The student voted in the most recent election *given* the student is part time
    d) Are the events voted and part time student independent or not? Explain.

5. There are 12 students in the dorm who are taking the same course. They want to form study groups of 4 students each. How many different such study groups can be formed from among the 12 students?

6. The Army gives a battery of exams to al new recruits. One exam measures a person's ability to work with technical machinery. The exams was given to a random sample of 360 new recruits. The results are shown below, where x is the score on a 10 point scale and f is the frequency of new recruits with this score.

x	1	2	3	4	5	6	7	8	9	10
f	28	42	79	83	51	36	18	12	7	4

a} If a person is chosen at random from those taking the test, use the relative frequency of test scores to estimate P(x) as x goes from 1 to 10.

b) Use a histogram to graph the probability distribution of part (a).

c) The ground-to-air missile battalion needs people with a score of seven or higher on the exam. What is the probability that a new recruit will meet this requirement?

d) The kitchen battalion can use people with a score of three or less. What is the probability that a new recruit is not overqualified for this work?

e) Compute the expected value μ of exam scores.

f) Compute the standard deviation of exam scores.

Chapter Test 5A

1. The New Mexico State Tourism and Travel Division Conducted a study in 1990 that showed 65% of the people who seek information about the state actually come for a visit. If the office receives 15 requests for information on the state, find the probabilities that, of those 15 people requesting information,
   a) exactly 15 people come visit the state.
   b) at least nine come visit the state.
   c) no more than eight come visit the state.
   d) from four to ten come visit the state (including four and ten)

2. A TV preference survey showed that 35% of all households watching evening cable TV are watching a sports-related program. Suppose a random sample of six households watching evening cable TV are contacted.
   a) Make a histogram showing the probability of r = 0, 1, 2, 3, 4, 5, 6 households watching a sports-related program.
   b) Find the means μ of this probability distribution. How many of the households do we expect to be watching a sports-related program?
   c) Find the standard deviation σ of the probability distribution.

3. Over the past several years Ms. Carver has determined that the probability of a successful sales call is 20%. What is the minimal number of sales calls Ms. Carver must make to be at least 89% sure of making at least one sale?

4. The probability that a customer will not be put on hold before reaching an airlines reservation person is 60%. Let n = 1, 2, 3, ... represent the number of times a person calls until the *first* time they are not put on hold.
   a) Write out a formula for the probability of the random variable n.
   b) What is the probability that the second time a person calls, he or she will reach a reservation person immediately (without being put on hold)?
   c) What is the probability a person needs to call more than 2 times to reach a reservation person directly (without being put on hold)?

5. Suppose the average number of customers entering a bank in a 30-minute period is 8. The bank wants to determine the probability that exactly 10 customers enter the bank in a 30 minute period.
   a) What is the value of λ?
   b) What is the probability that exactly 10 customers enter the bank during a 30-minute period?

6. The probability a new medication produces a bad side effect is 0.02. Estimate the probability that exactly 3 out of 100 patients will experience the bad side effect?

Chapter Test 5B

1. A survey by the National Occupational Information Coordinating Committee reports that 20% of employed adults expect to change jobs voluntarily in the next three years. If a random sample of 10 employed adults is selected, find the probability that in the next three years
   a) none of them expect to change jobs voluntarily.
   b) at least half of them expect to change jobs voluntarily.
   c) no more that half of them expect to change jobs voluntarily.
   d) from three to six (including three and six) expect to change jobs voluntarily.

2. An insurance company says 15% of all fires are caused by arson. A random sample of five fire insurance claims are under study.
   a) Make a histogram showing the probability of r = 0, 1, 2, 3, 4, 5 arson fires out of five fires.
   b) Find the mean μ of this probability distribution. What is the expected number of arson fires out of five fires?
   c) Find the standard deviation σ of the probability distribution.

3. Derrick scores on 70% of the free throws he attempts in basketball. What is the minimal number of free throws he must attempt to be at least 89% sure of making at least 2 of the free throws?

4. The probability of getting heads when you toss a fair coin is 0.50. Let n = 1, 2, 3, ... represent the number of times you toss a coin until the *first* head.
   a) Write out a formula for the probability of the random variable n.
   b) What is the probability that you must toss a coin 2 times in order to get heads?
   c) What is the probability you must toss the coin *more than* 2 times in order to get heads?

5. Suppose the average number of phone calls received by a business is 4 in a 15 minute period. The business wants to determine the probability that no calls will be received during a 15 minute period.
   a) What is the value of λ?
   b) Find the probability that no calls will be received during a 15 minute period.

6. On a typical day the probability one employee will call in sick is 0.04. Estimate the probability that exactly 5 out of 100 employees will call in sick tomorrow?

Chapter Test 6A

1. Let x represent the average miles per gallon of gasoline that owners get from their new Nippon model automobile. For this model care the mean of the x distribution is advertised to be μ = 44 mpg, with standard deviation σ = 6 mpg. Convert each of the following x intervals to standard z intervals.
   a) $x \geq 44$ b) $40 \leq x \leq 50$ c) $32 \leq x \leq 39$
   d) $x \leq 49$ e) $43 \leq x \leq 45$ f) $x \leq 38$

2. David and Laura are both applying for a position on a ski patrol team. David took the Advanced First Aid course at his college. His score on the comprehensive final exam was 173 points. The final exam scores followed a normal distribution with mean 150 and standard deviation 25. Laura took an Advanced First Aid course at the manufacturing plant where she works. For the method of testing used there, her cumulative score on all exams was 88. The cumulative scores followed a normal distribution with mean 65 and standard deviation 10. Both courses are comparable in content and difficulty. There is only one position available on the ski patrol and David and Laura are equally qualified as skiers.
   a) Both David and Laura scored 23 points above their respective means. Does the mean they both gave the same performance in the First Aid course? Explain you answer.
   b) Would you choose David or Laura for the ski patrol position based on knowledge of first aid and skiing ability? Explain your answer by locating both z scores for performance in the Fist Aid course under a standard normal curve.

3. Researchers at a pharmaceutical company have found that the effective time duration of a safe dosage of a pain relief drug is normally distributed with mean 2 hr and standard deviation 0.3 hr. For a patient selected at random, what is the probability that the drug will be effective.
   a) for 2 hours or less?
   b) for 1 hour or less?
   c) for 3 hours or more?

4. The life of a Freeze Breeze electric fan is normally distributed, with mean 4 years and standard deviation 1.2 years. The manufacturer will replace any defective fan free of charge while it is under guarantee. For how many years should a Freeze Breeze fan be guaranteed if the manufacturer does not want to replace more than 5% of them? (Give your answer to the nearest month.)

5. You are examining a quality control chart regarding number of employees absent each shift from a large manufacturing plant. The plant is staffed so that operations are still efficient when the average number of employees absent each shift is $\mu = 10.5$ with standard deviation $\sigma = 2$. For the most recent 12 shifts, the number of absent employees were

Shift	1	2	3	4	5	6	7	8	9	10	11	12
#	8	4	7	10	9	11	9	15	12	16	9	11

a) Make a control chart showing the number of employees absent during the 12 day period.

b) Are there any periods during which the number absent is out of control? Identify the out of control periods and comment about possible causes.

1. A high school counselor was given the following z intervals concerning a vocational training aptitude test. The test scores had a mean $\mu = 450$ and population standard deviation $\sigma = 35$ points. Convert each z interval to an x = test score interval.
   a) $-1.14 \leq z \leq 2.27$ b) $z \leq -2.58$ c) $1.645 \leq z$
   d) $-1.96 \leq z \leq 1.96$ e) $z \leq 0$ f) $1.28 \leq z \leq 1.44$

2. Jim scored 630 in a national bankers examination in which the mean is 600 and the standard deviation is 70. June scored 530 on the Hoople College banker's examination for which the mean is 500 and the standard deviation is 25. If Jim and June both apply for a job at Hoople State Bank and each examination has equal weight toward getting the job, who has the better chance? Explain your answer.

3. The ages of workers in the Illuminex plant are approximately normally distributed, with a mean of 45 years and a standard deviation of 12 years. A worker is stopped at random and asked to fill out a questionnaire. What is the probability that this worker is
   a) less than 30 years old?
   b) between 35 and 55 years old?
   c) more than 60 years old?

4. Quality control for Speedie Typewriters, Inc., has done studies showing that the light use model (150,000 words per year) has a normal distribution with a mean life of 3.5 years and a standard deviation of 0.7 years. At the prescribed usage, how long should the guarantee period be if the company wishes to replace no more than 10% during the guarantee period?

5. A toll free computer software support line for the spreadsheet Figure has established target length of time for each customer help phone call. The calls are targeted to have mean duration of 7 minutes with standard deviation 2 minutes. For one help technician the most recent 10 calls had the following duration.

call #	1	2	3	4	5	6	7	8	9	10
Length	8	15	10	4	6	4	8	12	10	12

   a) Make a control chart showing the lengths of calls
   b) Are there any periods during which the length of calls are out of control? Identify the out of control periods and comment about possible causes.

Chapter Test 7A

1. According to the Internal Revenue Service, the average tax refund for the 75 million personal income tax returns filed in 1988 was $875. Assuming a standard deviation of $230, find the probability that for a random sample of 36 returns the mean refund is
    a) less than $600.
    b) more than $1,000.
    c) between $600 and $1000.

2. True Sound cassette tapes have playing times that are *normally distributed*, with mean 30 min and standard deviation 2.3 min.
    a) What is the probability that a tape selected at random will play for a time period x between 28 min and 33 min?
    b) What is the probability that four tapes selected at random will have a mean playing time $\bar{x}$ between 28 min and 33 min?
    c) Compare your answers for parts (a) and (b). Was the probability in part (b) much higher? Why would you expect this?

3. Eureka Market Research Company conducts telephone interviews to determine product preference. They have found that the probability a person called at random agrees to answer a survey is 0.61. If 100 calls are made, what is the probability that
    a) 70 or more will respond to the survey?
    b) between 50 and 65 will respond?
    c) 55 or fewer will respond?

Chapter Test 7B

1. The Oak Grove College financial aid office did a study showing that their students spend an average (mean) of $680 in the college bookstore on books and supplies per year. The standard deviation is $138. If a random sample of 36 students is surveyed, what is the probability that the mean amount spent for books and supplies is
   a) less than $600?
   b) more than $700?
   c) between $600 and $700?

2. The diameters of grapefruit in a certain orchard are *normally distributed*, with mean 4.6 inches and standard deviation 1.3 inches. If a random sample of ten of these grapefruit are put in a bag and sold in a grocery store, what is the probability that the mean diameter $\bar{x}$ will be
   a) larger than 5 in.?
   b) between 4 in. and 5 in.?
   c) smaller than 4 in.?

3. The probability of an adverse reaction to a flu shot is 0.02. If the shot is given to 1,000 people selected at random, what is the probability that
   a) 15 or fewer people will have an adverse reaction?
   b) 25 or more people will have an adverse reaction?
   c) between 20 and 30 people will have an adverse reaction?

Chapter Test 8A

1. A random sample of 40 cups of coffee dispensed from an automatic vending machine showed that the mean amount of coffee the machine gave was $\bar{x} = 7.1$ oz with standard deviation $s = 0.3$ oz. Find a 90% confidence interval for the population mean of the amount of coffee dispensed.

2. In *Hospitals* (July, 1989) 18 economic forecasters made predictions about the average length of stay (i days) in hospitals for 1995. The mean of the predictions was 6.7 days, with standard deviation 0.5. Using this information, construct a 90% confidence interval for the predicted mean length of hospital stay in 1995.

3. A CPA is auditing the accounts of a large interstate banking system. Out of a random sample of 152 accounts it was found that 19 had transaction errors.
   a) Let p be the proportion of all such accounts with transaction errors. Find a point estimate for p.
   b) Find a 99% confidence interval for p.

4. An automobile manufacturer used a random sample of 50 cars of a certain model to estimate the miles per gallon (mpg) this model car gets in highway driving. The sample standard deviation was found to be 5.7 mpg. How many *more* cars should be included in the sample if we are to be 90% sure the sample mean $\bar{x}$ mpg is within 1 mpg of the population mean μ of all cars of this model?

5. The manager of Sam's Cookie Company is testing two processes for making chocolate chip cookies. A random sample of $n_1 = 90$ batches made with the first process had an average $\bar{x}_1 = 23$ broken cookies per batch with standard deviation $s_1 = 3$. Using the second process a random sample of $n_2 = 80$ batches had an average $\bar{x}_2 = 20$ broken cookies per batch with standard deviation $s_2 = 3.1$. Find a 90% confidence interval for the difference of average number of broken cookies made by the two processes. Does it appear (at the 90% confidence level) that the average number of broken cookies produced by the second process is less that than the average number produced by the first process? Explain.

6. Two pain relief drugs are being considered. A random sample of 6 doses of the first drug showed that the average amount of time required before the drug was absorbed into the blood stream was $\bar{x}_1 = 23$ min with standard deviation $s_1 = 4$ minutes. For the second drug, a random sample of 8 doses showed the average time required for absorption was $\bar{x}_2 = 27$ min with standard deviation $s_2 = 3.9$ minutes. Assume the absorption times follow a normal distribution. Find a 95% confidence interval for the difference in average absorption time for the two drugs. Does it appear that one drug is absorbed faster than the other (at the 95% confidence level)? Explain.

7. A random sample of 90 investment portfolios managed by Kendra showed that 75 of them met the targeted annual percent growth. A random sample of 110 investment portfolios managed by Lisa showed that 85 met the targeted annual percent growth. Find a 99% confidence interval for the difference in the proportion of the portfolios meeting target goals managed by Kendra compared to those managed by Lisa. Is there a difference in the proportions at the 99% confidence level? Explain.

Chapter Test 8B

1. Ralph Smith took a random sample of 40 homes in El Paso and found the homes received an average of 18.6 pieces of junk mail per week with a standard deviation of 5.2 pieces. Find a 0.95 confidence interval for the mean number of pieces of junk mail received per week by El Paso families.

2. Symptoms of a new flu virus have been determined from 18 sufferers to have a mean duration time of 9.7 days with standard deviation of 4.8 days. Find a 0.95 confidence interval for the mean duration time of these flu symptoms.

3. The Book Worm Book Club sends each of its members a book once a month. The members then have the option of returning the book without cost within 10 days. A random sample of 360 members showed that 216 had returned at least one book in the last year.
   a) Let p be the proportion of all members who returned at least one book last year. Find a point estimate for p.
   b) Find a 0.95 confidence interval for p.

4. At many gasoline stations customers have the option of paying cash and receiving a discount. The question is: What population of customers take advantage of this option?
   a) Let p be the proportion of customers who take advantage of the cash discount option. If no preliminary study is made to estimate p, how large a sample of customers is necessary to be 90% sure that a point estimate $\hat{p}$ will be within a distance of 0.08 of P?
   b) A preliminary study of 76 customers showed 19 used the cash discount option. How many *more* customers should be included in the sample to be 90% sure that a point estimate $\hat{p}$ will be within a distance of 0.08 of p?

5. A study of miles driven annual by U.S. households in the Midwest involved a random sample of size $n_1 = 15$ households for which the annual average miles driven was $\bar{x}_1 = 16.3$ thousand miles with standard deviation $s_1 = 4.1$ thousand miles. A random sample of $n_2 = 22$ households in the south showed the average number of miles driven annually was $\bar{x}_2 = 17.5$ thousand miles with standard deviation $s_2 = 4.2$ thousand miles. Assume the annual number of miles driven per household is approximately normally distributed. Find a 95% confidence interval for the difference in annual miles driven. At the 95% confidence level, is there a difference in average annual miles driven by households in the two regions? Explain.

6. A random sample of $n_1 = 35$ independent auto repair shops showed the average cost to make a specified repair was $\bar{x}_1 = \$740$ with standard deviation $s_1 = \$110$. A random sample of $n_2 = 40$ authorized dealers showed the average cost to make the same specified repair was $\bar{x}_2 = \$790$ with standard deviation $s_2 = \$50$. Find a 90% confidence interval for the difference in average repair charges between independent shops and dealers. At the 95% confidence level, is there a difference? Explain.

7. A study of voting patterns showed that from a random sample of 120 registered voters living in the inner city 55 voted in the most recent national election. A random sample of 200 registered voters from a rural area showed that 110 voted in the most recent national election. Find a 95% confidence interval for the difference in proportion of the registered voters who voted from the two types of regions. At the 95% confidence level, does there appear to be a difference? Explain.

Chapter Test 9A

For each of the following problems please provide the requested information.
- a) State the null and alternate hypotheses.
- b) Identify the sampling distribution to be used: the standard normal or the Student's t. Find the critical value(s).
- c) Sketch the critical region and show the critical value(s) on the sketch.
- d) Compute the z or t value of the sample test statistic and show it's location on the sketch of part (c).
- e) Find the P value or an interval containing the P value for the sample test statistic.
- f) Based on your answers for parts (a) to (e), shall we reject or fail to reject the null hypothesis? Explain your conclusion in simple nontechnical terms.

1. A large waterfront warehouse installed a new security system a few years ago. Under the old system the warehouse managers estimated that they were losing an average of $678 worth of merchandise to thieves each week. A random sample of 30 weekly records under the current system showed that they were still losing an average of $\bar{x}$ = $650 worth of merchandise each week. The sample standard deviation was s = $93. Does this indicate that the average loss each week is different (either more or less) than the previous $678 per week? Use a 5% level of significance.

2. *USA Today* reported that the average outstanding balance on regular Visa credit cards is $1113. A random sample of twelve Visa card holders who have had their accounts for only six months showed the average outstanding balance to be $\bar{x}$ = $750 with standard deviation $414. Use a 1% level of significance to test the claim that the people who have had the card for only six months have an average outstanding balance that is less than the national average.

3. The American Southpaw Association claims that the proportion of left-handed people in *Who's Who* is higher than the overall national proportion, which is one out of 21. A random sample of 180 people listed in the current *Who's Who* showed that 12 were left-handed. Is the claim justified at the 5% level of significance?

4. A test of 80 youths and 120 adults showed that 18 of the youths and 10 of the adults were careless drivers. Use a 1% level of significance to test the claim that the youth percentage of careless drivers is higher than the adult percentage.

5. A systems specialist has studied the work flow of clerks all doing the same inventory work. Based on this study, she designed a new work flow layout for the inventory system. To compare average production for the old and new methods, a random sample of six clerks was used. The average production rate (number of inventory items processed) for each clerk was measured both before and after the new system was introduced. The results are shown below. Test the claim that the new system speeds up the work rate (use $\alpha = 0.05$).

Clerk	1	2	3	4	5	6
B: Old M	110	100	97	85	117	101
A: New M	118	112	115	83	125	109

6. To test the claim that fluoride helps prevent cavities, 16 children had their teeth cleaned and coated with a fluoride emulsion. Another 16 children had their teeth cleaned, but received no fluoride coating. One year later the children without fluoride coating had a mean number of cavities $\bar{x}_1 = 2.4$ with sample standard deviation $s_1 = 0.8$. for the children with the fluoride coating the mean was $\bar{x}_2 = 1.7$ with $s_2 = 1.0$. Is the claim justified at the 5% level of significance?

7. A large company has been hiring graduates from two secretarial schools. The company efficiency expert gave a typing test to a random sample of 30 new graduates from one school and found the graduates' mean score to be $\bar{x}_1 = 68$ words per minute with standard deviation $s_1 = 16$. Another random sample of 30 new graduates from the other school was tested. The mean score of the second group was $\bar{x}_2 = 71$ words per minute with standard deviation $s_2 = 12$. At the 5% level of significance, can we say there is a significant difference between the average scores?

Chapter Test 9B

For each of the following problems please provide the requested information.
  a) State the null and alternate hypotheses.
  b) Identify the sampling distribution to be used: the standard normal or the Student's t. Find the critical value(s).
  c) Sketch the critical region and show the critical value(s) on the sketch.
  d) Compute the z or t value of the sample test statistic and show it's location on the sketch of part (c).
  e) Find the P value or an interval containing the P value for the sample test statistic.
  f) Based on your answers for parts (a) to (e), shall we reject or fail to reject the null hypothesis?   Explain your conclusion in simple nontechnical terms.

1. A new state-run job assistance program helps the unemployed find work. In one community where there is a job assistance program, a sociologist is studying unemployment. The sociologist found that before the job assistance program, the mean length of time it took an unemployed person to find work was 57 days. A random sample of 33 people using the new job assistance program showed that it took an average of 48 days to find work. The sample standard deviation was 14 days. Does this indicate that the average length of time to find a job is now less than 57 days? Use a 5% level of significance.

2. The airport has made improvements in its baggage claim system. It used to take passengers a mean of 20 min to get their luggage. After the improvements were made, a random sample of 14 passengers received their luggage in an average of $\bar{x}$ = 18.6 min with standard deviation 2.8 min. Use a 1% level of significance to test if the meantime required for the passengers to receive their baggage is different (either shorter or longer) from 20 min.

3. Internal Revenue Service employees who process income tax returns say that 10% of all tax returns contain arithmetic errors in excess of $1,200. A random sample of 817 income tax returns showed that 86 of them contain such errors. Do these data indicate that the proportion of income tax returns with arithmetic errors in excess of $1200 is different from 10%? Use a 1% level of significance.

4. A random sample of 530 union bricklayers showed that 40 were unemployed. A random sample of 640 nonunion bricklayers showed that 60 were unemployed. Do these data indicate that the proportion of unemployed bricklayers is greater for the nonunion people? Use a 5% level of significance.

5. Five members of the college track team in Denver (elevation 5,200 ft) went up to Leadville (elevation 10,152 ft) for a track meet. The times in minutes for these team members to run two miles at each location are shown in below

Team Member	1	2	3	4	5
Denver	10.3	9.8	11.4	9.7	9.2
Leadville	11.5	10.6	11.0	10.8	10.1

   Assume the team members constitute a random sample of track team members. Use a 5% level of significance to test the claim that the times were longer at the higher elevation.

6. Linda is an Alpine botanist who thinks she has discovered a new species of wildflower. The only morphological difference from that of a known species is the petal length. A random sample of 12 flowers of the known species has a mean petal length $\bar{x}_1$ = 9.3 mm and sample standard deviation s_1 = 1.1 mm. A random sample of 15 "new species" flowers has mean petal length $\bar{x}_2$ = 11.9 mm with sample standard deviation s_2 = 1.9 mm. Linda claims that the mean petal lengths of the two types of flowers are different. Is this claim justified at the 1% level of significance?

7. Inner city residents claim that their average grocery costs are higher than the average grocery costs of people living in the suburbs. To test this claim the price of a quart of milk, a pound of bread, a dozen medium eggs, and a pound of butter was obtained from a random sample of 30 suburban stores. The average cost was $5.20 with standard deviation s_1 = $0.30. For a random sample of 36 inner city stores the same items cost an average of $\bar{x}_2$ = $5.54 with standard deviation s_2 = $0.50. Is the claim justified at the 5% level of significance?

Chapter Test 10A

Urban travel times and distances are important factors in the analysis of traffic flow patterns. A traffic engineer in Los Angeles obtained the following data from area freeways, where x = miles traveled and y = time in minutes for a passenger car.

x (miles)	5	9	3	11	20	15	12	25
y (min)	9	13	6	28	21	21	16	31

1. Draw a scatter diagram.
2. Find the equation of the least squares line and graph the least squares line on the scatter diagram of problem 1
3. Find the value of the correlation coefficient r and the coefficient of determination r^2, Give a brief explanation of the coefficient of determination in the context of this problem.
4. Find the standard error of estimate S_e.
5. If a passenger car travels 20 miles on the Los Angeles freeway, what is the predicted length of time in minutes as predicted by the least squares line?
6. Find a 90% confidence interval for your prediction in problem 5.
7. Is r (computed in problem 3) significant at the $\alpha = 0.01$ level of significance?

Chapter Test 10B

The following data are taken from *Fortune* magazine. The data represents five aerospace companies (chosen at random) from the Fortune listing of aerospace companies. In this data, x = number of company employees (in ten thousands) and y = annual profits (in hundreds of millions of dollars).

x	11.2	8.7	6.8	3.2	4.5
y	8.1	6.2	3.6	0.9	1.0

1. Draw a scatter diagram.
2. Find the equation of the least squares line and graph the least squares line on the scatter diagram of problem 1.
3. Find the value of the correlation coefficient r and the coefficient of determination r^2, Give a brief explanation of the coefficient of determination in the context of this problem.
4. Find the standard error of estimate S_e.
5. If an aerospace company employs 5 (in ten thousands) employees, find the annual profits (in hundreds of millions of dollars) as predicted by the least squares line.
6. Find a 90% confidence interval for your prediction in problem 5.
7. Is the correlation coefficient r computed in problem 3 significant at the 5% level of significance?

Chapter Test 11A

1. The management of a large corporation is conducting a study of employee satisfaction under two work schedules: a 4 day work week with 10 hour shifts per day and a 5 day work week with 8 hour shifts per day.   A random sample of 300 employees were assigned one of the two work schedules. After six months on the schedule, these employees were surveyed. The results follow.

Satisfaction	4 day schedule	5 day schedule	Row Total
High	80	70	150
Medium	30	40	70
Low	40	40	80
Column Total	150	150	300

Use a chi-square test to determine if work schedule and satisfaction level are independent at the 0.05 level of significance.

2. Last year the tourist industry reported that 25% of the package tours sold were of duration less than 4 days, 60% of the trips lasted between 4 and 10 days, and 15% lasted more than 10 days. This year a random sample of 200 vacation packages sold showed that 70 were for trips lasting less than 4 days, 110 were for trips lasting between 4 and 10 days, and 20 were for trips lasting more than 10 days. Are package tours sold this year of different duration than those sold last year? Use a 0.05 level of significance.

3. The variance on the duration of one dose of pain medication is specified to be $\sigma^2 = 15$ minutes. A random sample of 30 patients were given the medication and the sample standard deviation of the duration of the medication was $s^2 = 18$ minutes. Is this evidence that the variance of the duration of the medication is too large? Use a 1% level of significance.

4. A study to determine if management style affects the number of sick leave days taken by employees in a department was conducted. Three departments with the same number of employees were studied. The management style used in one department was top down with employees having little input into decisions; in another department formal quality teams made recommendations; in the last department the management gathered input informally from the employees. The total number of sick leave days taken per month by all of the employees in the department was recorded. For a random sample of 3 months, the numbers follow:

Top down management: 15 25 19
Quality teams: 18 22 16
Informal input: 19 13 27

Use ANOVA to test if the mean number of sick leave days for departments managed in the various styles are different. Use $\alpha = 0.05$.

Chapter Test 11B

1. Highlands State College is doing a study to determine if fees for course schedule changes have any effect on the number of course schedule changes students make during the drop/add period. A random sample of student schedules showed the following data.

Schedule	No fee	$25 fee	Row Total
No changes	125	135	260
Changes	75	65	140
Column Total	200	200	400

   Use a 1% level of significance to test the claim that the number of course schedule changes is independent of the fee charged.

2. Has the length of telephone calls changed over the past ten years? Ten years ago 10% of the phone calls lasted less than one minute, 60% lasted between 1 and 10 minutes, and 30% lasted longer than 10 minutes. This year, a random sample of 1000 calls showed that 250 lasted less than I minute; 500 lasted between 1 and 10 minutes, and 250 lasted longer than 10 minutes. Use a 1% level of significance to test the claim that duration of phone calls is different now than it was ten years ago.

3. The manager of an electronic parts supplier is tracking the variance in the time between a parts order and delivery of the part. Company standards specify that the variance should be no more than $\sigma^2 = 2$ days. A random sample of 41 orders showed the sample variance to be $s^2 = 2.8$ days. At the 5% level of significance, does the variance seem larger than that specified?

4. A study of depression and exercise was conducted. Three groups were used: those in a designed exercise program; a group that is sedentary; and a group of runners. A depression rating (higher scores meaning more depression) was given to the participants in each group. Small random samples from each groups provided the following data on the depression rating:

   Treatment Group: 52 58 47
   Sedentary Group: 57 50 60
   Runners: 51 49 60

   Use ANOVA to test if the mean depression ratings for the three groups are different. Use $\alpha = 0.05$.

Chapter Test 12A

1. The management of a large retail store did a study of sales in different departments last year and this year and ranked the departments by sales volume (with lowest rank meaning highest sales). The data follow.

Department	1	2	3	4	5	6	7	8
Last year rank	1	7	4	5	3	2	6	8
This year rank	1	6	5	7	2	3	4	8

   Test the claim at the 0.01 level of significance that there is a monotone relation either way between last year's and this year's performance.

2. A workshop on harmony in the work place was given to a randomly selected group of employees. Another group did not participate in the workshop. A test measuring sensitivity to other viewpoints was given to both groups with higher scores indicating more sensitivity. The results follow.

Workshop participant	73	81	91	56	78	83	52	92
Non-workshop	85	70	74	55	90	48	75	86

   Test the claim at the 5% level of significance that there is a difference either way in the average sensitivity score for the two groups.

3. A restaurant wants to determine if a newspaper coupon giving diners a free desert increases the number of diners per night. The number of diners during the week before the coupon became effective and the number of diners each night during the week the coupon was in effect are given below.

	Mon	Tue	Wed	Thur	Fri	Sat	Sun
No Coupon	73	53	65	84	112	140	97
Coupon	82	47	70	65	120	130	160

   Use a 5% level of significance to test the hypothesis that the mean number of diners coming into the restaurant when the coupon was in effect was higher.

Chapter Test 12B

1. Professor Smith gives a midterm and final exam. For a random sample of ten students, the class rank on the two exams follow, with a lower rank number meaning a higher score.

Student	1	2	3	4	5	6	7	8	9	10
Midterm	5	3	1	7	10	2	8	4	6	9
Final	9	4	5	7	8	3	6	1	2	10

Test the claim that there is a monotone increasing relation between the ranks of the two exams. Use a 5% level of significance.

2. A random sample of households with an income level below $30,000 were asked to record the number of hours per week someone in the household was watching T.V. A random sample of households with income level at or above $30,000 were asked to record the same information. The results follow.

Less than $30,000	100	40	35	70	80	90	15	75
$30,000 or more	85	62	41	37	45	91	30	10

Test the claim at the 5% level of significance that there is a difference either way in the average number of hours households in the two income categories watch T.V.

3. A self confidence inventory instrument was administered to a group of students before and after a self confidence training workshop. The scores follow with a higher score indicating more self confidence.

Student	1	2	3	4	5	6	7	8	9
Before	35	42	37	45	43	47	33	37	35
After	40	38	37	43	45	46	41	40	36

Use a 5% level of significance to test the hypothesis that the mean scores were higher after the workshop.

Solutions to Chapter Tests

Chapter Test 1A

1. a) Number of hours of sleep for all college students enrolled for 15 credits on Tuesday nights
   b) Number of hours of sleep on Tuesday nights for each of the 50 students included in the sample
   c) No, hours of sleep on Friday night might be different than hours of sleep on Tuesday night. Also the credit hour load of the students needs to be 15

2. a) ratio b) interval c) nominal d) ordinal c) ordinal

3. a) census b) experiment c) sampling d) simulation

4. *see* text

Chapter Test 1B

1. a) Observation results regarding illegal left turn of all cars (past and future) who exit the post office parking lot under the present law enforcement patterns..
   b) Observed results regarding illegal left turn of the 200 cars included in the sample. Our data values might be "yes" for an illegal left turn and "no" for a car not making an illegal left turn.
   c) No, these observations apply only to the cars exiting the post office parking lot under study.

2. a) nominal b) ratio c) interval d) ordinal e) ratio

3. a) census b) sample c) simulation d) experiment

4. *see* text

Chapter Test 2A

1. The outcomes are the totals 2 through 12. Group digits of the random number table in groups of 2. Record permissible outcomes, allowing repetition, until you have 4 outcomes. The first outcome is 9.

2.

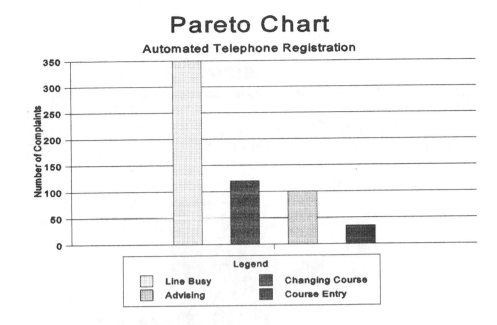

Pareto Chart
Automated Telephone Registration

Legend
- Line Busy
- Advising
- Changing Course
- Course Entry

3. a) CW = 5

Class Boundary	Frequency	Relative Frequency	Cumulative Frequency
0.5 - 5.5	12	0.48	12
5.5 - 10.5	4	0.16	16
10.5 - 15.5	7	0.28	23
15.5 - 20.5	1	0.04	24
20.5 - 25.5	1	0.04	25

3. continued
   b)

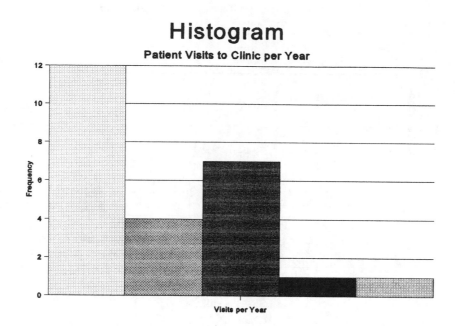

Histogram

Patient Visits to Clinic per Year

c) Relative frequency histogram has same shape, but each entry on the vertical axis is divided by sample size of 25.
d)

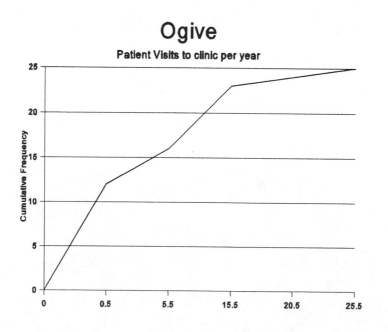

Ogive

Patient Visits to clinic per year

Chapter Test 2A continued

4. <u>Price of Tickets from Scalpers</u>
        unit = $1
        1| 6 represents $16

        1 | 6
        2 | 0 4 5 5 5 7 8
        3 | 0 0 1 3 3 5 5 5 5 5 5 6 7 7 8 9
        4 | 0 0 0 5 5
        5 | 0

5. a) Systematic   b) Cluster   c) Stratified (strata are the age groups)
    d) Convenience   e) Random

6.

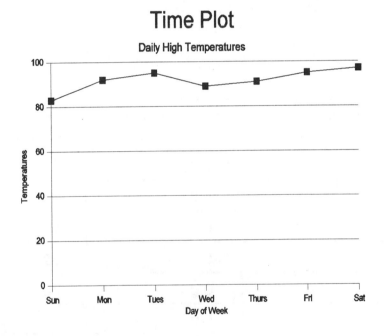

Chapter Test 2B

1. Number the employees with distinct numbers starting with number 1 and going to number 736. Group the digits in the random number table into groups of three. Use the groups of digits to select 30 distinct members for the sample. The numbers of the first five members of the sample are 718, 645, 036, 507, 093

2.

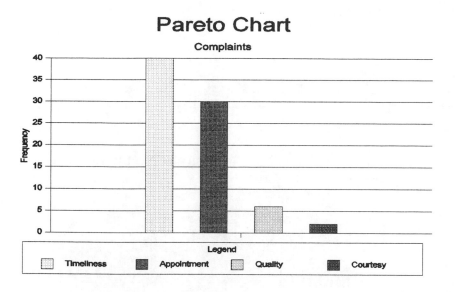

3. a) Class width = 6

Class Boundary	Frequency	Relative Frequency	Cumulative Frequency
1.5 - 7.5	7	0.28	7
7.5 - 13.5	6	0.24	13
13.5 - 19.5	6	0.24	19
19.5 - 25.5	3	0.12	22
25.5 - 31.5	2	0.08	24
31.5 - 37.5	1	0.04	25

3 . b)

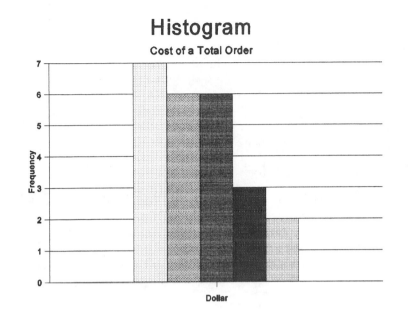

Histogram

Cost of a Total Order

c)

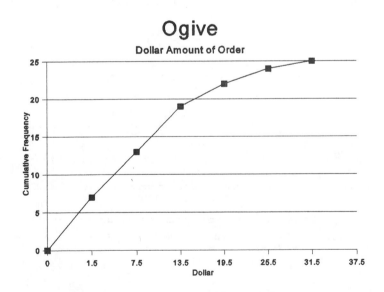

Ogive

Dollar Amount of Order

d) The relative frequency histogram will have the same shape as the histogram. However, the y axis will have relative frequencies obtained by dividing each y value by 25.

4.   Time Elapsed
        unit = I day
        1| 6 represents 26 days

2	1 1 2 6 7
3	2 3 7
4	2 3 5 7 8 8 9
5	4 7 8
6	3 7

5. a) Random b) Stratified (strata are the number of bedrooms) c) Cluster
    d) Convenience   e) Systematic

6.

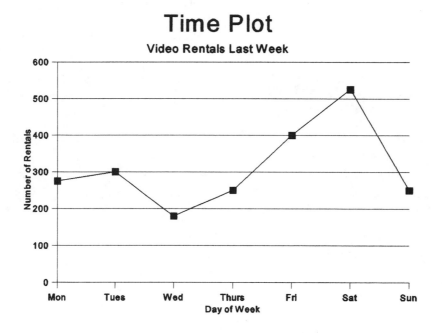

Chapter Test 3A

1. Mean = 7.625; Median = 8; Mode = 8

2. a) Range = 5 hr   b) $\bar{x}$ = 3.8 hr   c) s = 1.92 hr

3. a) CV = 11.90   b) Interval is from $\bar{x}$ - 2σ to $\bar{x}$ + 2σ or from 32 to 52

4. a) $\bar{x} \approx$ 27.26   b) s $\approx$ 10.32

5. a) low value = 19, Q_1 = 27, median = 32.5, Q_3 = 41, high value = 55
   b)

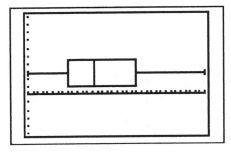

6. 83.75

Chapter Test 3B

1. Mean = 32.35;   Median = 26.5;   Mode = 21

2. a) Range = 87.80 b) $\bar{x}$ = 61.56   c) s = 31.21

3. a) CV = 16.22   b) Interval from $\bar{x}$ + 2σ to $\bar{x}$ - 2σ or 25 to 49

4. a) $\bar{x}$ = $280.4   b) s = $108.5

5. a) low value = 15, Q_1 = 31, median = 39.5, Q_2 = 52, high value = 72
   b)

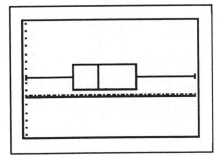

6. 3.92

Chapter Test 4A

1. a) 0.06   b) 0.94   c) defective, not defective; yes

2. a) 10/64   b) 10/56

3. P(Pass written *and* pass physical) = (0.82)(0.58) = 0.48

4. a) 50/70   b) 25/70   c) 25/30
   d) Not independent since the probabilities in part (a) and (c) are different

5. 6 phones

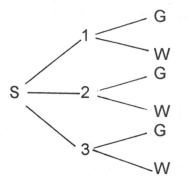

6. a)

x	0	1	2	3	4	5	6	7	8	9	10
P(x)	.162	.196	.242	.146	.073	.062	.046	.038	.023	.008	.004

b)

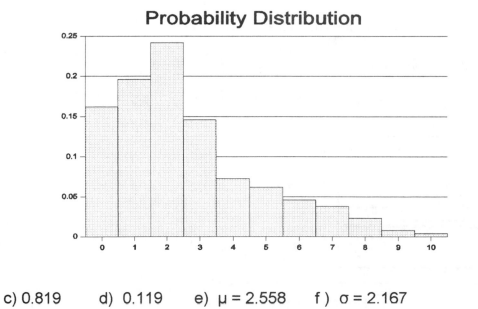

c) 0.819   d) 0.119   e) $\mu = 2.558$ f) $\sigma = 2.167$

Chapter Test 4B

1. a) 3/5   b) 2/5   c) 1/5

2. a) 8/81   b) 8/72

3. P(woman *and* engineering major) = (0.53)(0.20) = 0.1060

4. a) 35/90   b) 15/90   c) 15/40
   d) Not independent since the probabilities in part (a) and (c) are different

5. $C_{12,4} = 495$

6. a)

x	1	2	3	4	5	6	7	8	9	10
P(x)	.078	.117	.219	.231	.142	.100	.050	.033	.019	.011

b)

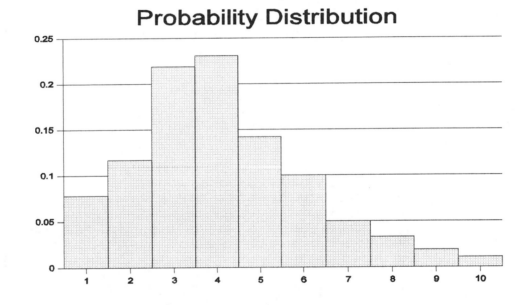

Probability Distribution

c) 0.113   d) 0.414   e) $\mu = 4.098$ f) $\sigma = 1.942$

Chapter Test 5A

1. p = 0.65 and n = 15
   a) P(r = 15) = 0.002   b) P(r $\geq$ 9) = 0.756
   c) p(r $\leq$ 8) = 0.245   c) P(4 $\leq$ r $\leq$ 10) = 0.648

2. a)

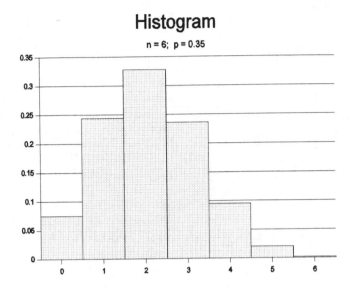

   b) μ = 2.10
   c) σ = 1.17

3. n = 10

4. a) P(n) = 0.60(0.40)$^{n-1}$ b) 0.24   c) 0.16

5. a) λ = 8   b) 0.0993

6. 0.1804

Chapter Test 5B

1. p = 0.20 and n = 10
   a) P(r = 0) = 0.107   b) P(r $\geq$ 5) = 0.033
   c) P(r $\leq$ 5) = 0.992   d) P(3 $\leq$ r $\leq$ 6) = 0.321

2. a)

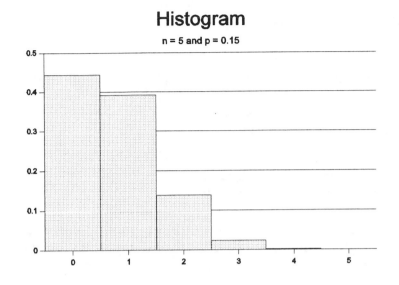

Histogram
n = 5 and p = 0.15

b) μ = 0.750
c) σ = 0.798

3. n = 4

4. a) $P(n) = 0.5(0.5)^{n-1}$ b)   0.25   c) 0.25

5. a) λ = 4   b) 0.0183

6. 0.1563

Chapter Test 6A

1. a) $z \geq 0$ b) $-0.67 \leq z \leq 1.00$ c) $-2.00 \leq z \leq -0.83$
   d) $z \leq 0.83$ e) $-0.17 \leq z \leq 0.17$ f) $z \leq -1.00$

2. a) No   b) For David, z = 0.92; For Laura, z = 2.30; Choose Laura

3. a) 0.5000   b) 0.0004   c) 0.0004

4. 2 years or 24 months

5. a)

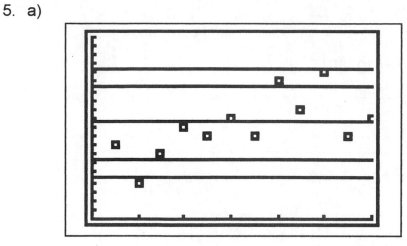

b) Day 2 is out of control since one value is lower than three standard deviations below the mean. Days 8, 9, and 10 are out of control because 2 out of 3 points are beyond 2σ. Explanations vary.

Chapter Test 6B

1. a) $410.1 \leq x \leq 529.5$ b) $x \leq 359.7$ c) $507.6 \leq x$
   d) $381.4 \leq x \leq 518.6$ e) $x \leq 450$ f) $494.8 \leq x \leq 500.4$

2. For Jim $z = 0.43$; for June $z = 1.20$; Choose June

3. a) 0.1056   b) 0.5934   c) 0.1056
4. 2.6 yr
5. a)

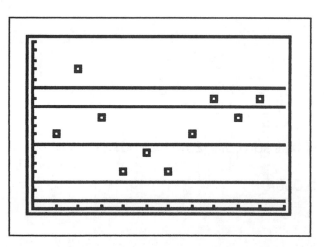

b) Day 2 is out of control because the length of the call is more than 3σ above the mean. Days 8, 9, and 10 are out of control because 2 of these three days are more than 2σ above the mean. Explanations vary.

Chapter Test 7A

1. a) $P(\bar{x} < 600) = P(z < -7.17) \approx 0.0001$
   b) $P(\bar{x} > 1000) = P(z > 3.26) = 0.0006$
   c) $P(600 < \bar{x} < 1000) = P(-7.17 < z < 3.26) \approx 0.9993$

2. a) 0.7110   b) 0.9546   c) Yes, the standard deviation in part (b) is smaller

3. a) $P(r \geq 70) = P(x \geq 69.5) = P(z \geq 1.74) = 0.0409$
   b) $P(50 \leq r \leq 65) = P(49.5 \leq x \leq 65.5) = P(-2.36 \leq z \leq 0.92) = 0.8121$
   c) $P(r \leq 55) = P(x \leq 55.5) = P(z \leq -1.13) = 0.1292$

Chapter Test 7B

1. a) $P(\bar{x} < 600) = P(z \leq -3.48) = 0.0003$
   b) $P(\bar{x} \geq 700) = P(z \geq 0.87) = 0.1922$
   c) $P(600 \leq \bar{x} \leq 700) = P(-3.48 \leq z \leq 0.87) = 0.8075$

2. a) $P(\bar{x} \geq 5) = P(z \geq 0.97) = 0.1660$
   b) $P(4 \leq \bar{x} \leq 5) = P(-1.46 \leq z \leq 0.97) = 0.7619$
   c) $P(\bar{x} \leq 4) = P(z \leq -1.46) = 0.0721$

3. a) $P(r \leq 15) = P(x \leq 15.5) = P(z \leq -1.02) = 0.1539$
   b) $P(r \geq 25) = P(x \geq 24.5) = P(z \geq 1.02) = 0.1539$
   c) $P(20 \leq r \leq 30) = P(19.5 \leq x \leq 30.5) = P(-0.11 \leq z \leq 2.37) = 0.5349$

Chapter Test 8A

1. 7.02 to 7.18

2. 6.5 to 6.9

3. a) $\hat{p} = 0.125$ b) 0.06 to 0.19

4. 38 more

5. $\bar{x}_1 - \bar{x}_2 = 3$; The confidence interval for $\mu_1 - \mu_2$ is from 2.23 to 3.77; The confidence interval contains numbers that are all positive so it appears the average number of broken cookies under process one is greater than the average number of broken cookies under the second process.

6. $\bar{x}_1 - \bar{x}_2 = -4$; $s = 3.942$; d.f. = 12; $t_{0.95} = 2.179$; The confidence interval for $\mu_1 - \mu_2$ is from -8.64 to 0.64; Since the confidence interval includes both positive and negative numbers, we cannot conclude there is a difference in average absorption times at the 95% confidence level.

Chapter Test 8A

7. Let $\hat{p}$ be the sample proportion of successful portfolios managed by Kendra and $\hat{p}_2$ be the sample proportion managed by Lisa. $\hat{p}_1 - \hat{p}_2 \approx 0.0606$; The confidence interval for $p_1 - p_2$ is from -0.084 to 0.205; Since the confidence interval contains both positive and negative numbers we conclude there is no difference in proportions at the 99% confidence level.

Chapter Test 8B

1. 16.99 to 20.21

2. 7.31 to 12.09

3. a) $\hat{p} = 0.6$ b) 0.55 to 0.65

4. a) 106   b) 4 more

5. $\bar{x}_1 - \bar{x}_2 = -1.2$; d.f. = 35; $t_{0.95} = 2.030$; s = 4.16; The confidence interval for $\mu_1 - \mu_2$ is from -4.03 to 1.63; Since the confidence interval contains both positive and negative numbers we conclude at the 95% confidence level that there is no difference in the means.

6. $\bar{x}_1 - \bar{x}_2 = -50$; The confidence interval for $\mu_1 - \mu_2$ is from -10.4 to -89.6; Since the confidence interval contains only negative numbers we conclude that at the 95% confidence level, the average amount charged by independent shops is less than the average amount charged by the dealers.

7. Let $\hat{p}_1$ be the sample proportion of voters from the inner city and $\hat{p}_2$ be the sample proportion of voters from the rural area; $\hat{p}_1 - \hat{p}_2 = -0.916$; The confidence interval for $p_1 - p_2$ is from -0.20 to 0.02; Since the confidence interval contains both positive and negative numbers, we conclude at the 95% confidence level that there is no difference in the proportion of registered voters who voted in inner city and rural area.

Chapter Test 9A

1. H_0: $\mu = 678$ H_1: $\mu \neq 678$; $z_0 = \pm 1.96$; sample test statistic $\bar{x} = 650$ corresponds to z = -1.65; P value = 0.0990; Do not reject H_0; There is not enough evidence to say the average loss has changed.

2. H_0: $\mu = 1113$ H_1: $\mu < 1113$; d.f. = 11; $t_0 = -2.718$; sample test statistic $\bar{x} = 750$ corresponds to t = -3.037; 0.005 < P value < 0.10; Reject H_0; The evidence indicates that card holders of only six months have a lower average balance.

Chapter Test 9A

3. H_0: $p = 1/21$ H_1: $p > 1/21$; $z_0 = 1.645$; sample test statistic $\hat{p} = 12/180$ corresponds to $z = 1.20$; P value = 0.1151; Do not reject H_0; The proportion of left-handers in *Who's Who* does not appear to be greater.

4. Let p_1 represent the proportion of youth who are careless drivers and p_2 the proportion of adults who are careless drivers. H_0: $p_1 = p_2$ H_1: $p_1 > p_2$; $z_0 = 2.33$; sample test statistic $\hat{p}_1 - \hat{p}_2 = 0.1417$ corresponds to $z = 2.83$; P value = 0.0023; Reject H_0; The proportion of youth careless drivers seems to be higher.

5. H_0: $\mu_d = 0$ H_1: $\mu_d < 0$; d.f. = 5; $t_0 = -2.015$; sample test statistic $\bar{d} = -8.67$ corresponds to $t = -3.127$; 0.01 < P value < 0.025; Reject H_0; At the 5% level of significance the new system speeds up the work rate.

6. H_0: $\mu_1 = \mu_2$ H_1: $\mu_1 > \mu_2$; d.f. = 30; $t_0 = 1.697$; s = 0.9055; sample test statistic $\bar{x}_1 - \bar{x}_2 = 0.7$ corresponds to $t = 2.187$; 0.010 < P value < 0.025; Reject H_0; Those treated with fluoride seem to have fewer cavities.

7.. H_0: $\mu_1 = \mu_2$ H_1: $\mu_1 \neq \mu_2$; $z_0 = \pm 1.96$; sample test statistic $\bar{x}_1 - \bar{x}_2 = -3$ corresponds to $z = -0.82$; P value = 0.4122; Do not reject H_0; The schools do not seem to differ.

Chapter Test 9B

1. H_0: $\mu = 57$; H_1: $\mu < 57$; $z_0 = -1.645$; sample test statistic $\bar{x} = 48$ corresponds to $z = -3.69$; P value ≈ 0.0001; Reject H_0; It seems the average length of time to find a job is now less than 57 days.

2. H_0: $\mu = 20$; H_1: $\mu \neq 20$; D.f. = 13; $t_0 = \pm 3.012$; sample test statistic $\bar{x} = 18.6$ corresponds to $t = -1.871$; 0.050 < P value < 0.100; Do not reject H_0; There is not enough evidence to conclude that the average waiting time for baggage is different from 20 minutes.

3. H_0: $p = 0.10$ H_1: $p \neq 0.10$; $z_0 = \pm 2.58$; sample test statistic $\hat{p} = 0.1053$ corresponds to $z = 0.50$; P value = 0.6170. Do not reject H_0; There is not enough evidence to say the percentage of returns containing the described error is different from 10%.

4. Let p_1 be the proportion from union workers and p_2 be the proportion from nonunion workers; H_0: $p_1 = p_2$; H_1: $p_1 < p_2$; $z_0 = -1.645$; $\bar{p} \approx 0.0855$; sample test statistic $\hat{p}_1 - \hat{p}_2 = -0.018$ corresponds to $z = -1.11$; P value = 0.1335; do not reject H_0; The unemployment rates among union and nonunion workers do not show any difference.

Chapter Test 9B

5. H_0: $\mu_d = 0$ H_1: $\mu_d < 0$; d.f. = 4; $t_0 = -2.132$; sample statistic $\bar{d} = -0.72$ corresponds to $t = -2.493$; $0.025 < P$ value < 0.05; Reject H_0; It appears the Leadville times are longer than the Denver times.

6. H_0: $\mu_1 = \mu_2$ H_1: $\mu_1 \neq \mu_2$; d.f. = 25; $t_0 = \pm 2.787$; $s = 1.598$; sample test statistic $\bar{x}_1 - \bar{x}_1 = -2.6$ corresponds to $t = -4.225$; P value < 0.010; Reject H_0; The sample evidence indicates that the petal lengths are different.

7. H_0: $\mu_1 = \mu_2$ H_1: $\mu_1 < \mu_2$; $z_0 = -1.645$; sample test statistic $\bar{x}_1 - \bar{x}_2 = -0.34$ corresponds to $z = -3.41$; P value $= 0.0003$; Reject H_0; The sample evidence indicates the inner city residents pay more for the specified groceries than the suburban residents.

Chapter Test 10A

1.

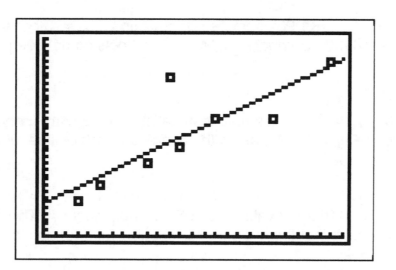

2. $y = 0.993x + 5.707$; *See* line in figure of problem 1

3. $r = 0.83$; $r^2 = 0.69$

4. $S_e = 5.26$ (answers will vary slightly depending on rounding)

5. for $x = 20$, $y = 25.58$ (answers will vary slightly depending on rounding)

6. 14.0 to 37.1 (answers will vary slightly depending on rounding)

7. H_0: $\rho = 0$ H_1: $\rho > 0$; critical value is 0.79; Since $r = 0.83$ lies in the critical region we reject H_0 and conclude that r is significant.

Chapter Test 10B

1.

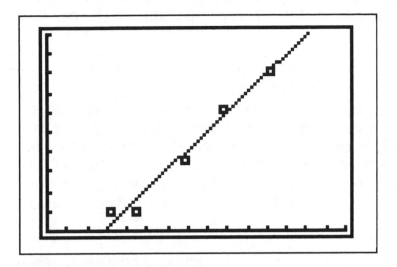

2. $y = 0.978x - 2.77$ (*See* line on figure in problem 1)

3. $r = 0.988$; $r^2 = 0.976$

4. $S_e = 0.574$ (Answers will vary slighly depending on rounding)

5. For $x = 5$, $y = 2.12$ (Answers will vary slightly depending on rounding)

6. 0.59 to 3.65 (Answers will vary slightly depending on rounding)

7. H_0: $\rho = 0$ H_1: $\rho > 0$; critical value is 0.81; Since $r = 0.988$ is in the critical region, we reject H_0 and conclude that r is significant.

Chapter Test 11A

1. H_0: Work schedules and employee satisfaction are independent; H_1: Work schedules and employee satisfaction are not independent; $\chi^2 = 2.095$; $\chi^2_{0.05} = 5.99$; Do not reject H_0.

2. H_0: The distribution of length of vacation package this year is the same as last year; H_1: The distribuitons are different; $\chi^2 = 12.17$; $\chi^2_{0.05} = 5.99$; Reject H_0; The distribuiton of trip lengths seems to have changed.

3. H_0: $\sigma^2 = 15$; H_1: $\sigma^2 > 15$; $\chi^2 = 34.8$; $\chi^2_{0.01} = 49.59$; Do not reject H_0; The variance seems to be as specified.

Chapter Test 11A

4. H_0: All the means are equal; H_1: Not all the means are equal; SS(BET) = 2; SS(W) = 168; SS(TOT) = 170; DF(BET) = 2; DF(W) = 6; DF(TOT) = 8; MS(BET) = 1; MS(W) = 28; F = 0.036; $F_{0.05}$ = 5.14; Do not reject H_0. The means do not seem to be different among the groups.

Chapter Test 11B

1. H_0: Number of schedule changes and fee are independent; H_1: Number of schedule changes and fee are not independent; χ^2 = 1.0989; $\chi^2_{0.01}$ = 6.63; Do not reject H_0; The number of changes seem to be independent of fee.

2. H_0: Distributions of call lengths is the same; H_1: Distributions of call lengths has changed; χ^2 = 250; $\chi^2_{0.01}$ = 9.21; Reject H_0; The distribution of length of calls seems to have changed.

3. H_0: σ^2 = 2; H_1: σ^2 > 2: χ^2 = 56; $\chi^2_{0.05}$ = 55.76; Reject H_0; The variance seems larger than specified.

4 H_0: All the means are equal; H_1: The means are not all equal; SS(BET) = 17.56; SS(W) = 182; SS(TOT) = 199.56; DF(BET) = 2; DF(W) = 6; DF(TOT) = 8 MS(BET) = 8.78; MS(W) = 30.33; F = 0.29; $F_{0.05}$ = 5.14; Do not reject H_0; The means do not seem to be different among the groups.

Chapter Test 12A

1. H_0: ρ_S = 0; H_1: $\rho_S \neq 0$; critical value = 0.881; Since the sample test statistic r_S = 0.857 does not fall in the critical region, we fail to reject H_0. There does not seem to be a monotone relation.

2. H_0: Workshop makes no difference; H_1: Workshop makes a difference; $\pm z_0 = \pm 1.96$; μ_R = 68; σ_R = 9.522; R(workshop participants) = 73 corresponds to z = 0.525; Since the sample test statistic does not fall in the critical region, we fail to reject H_0. The workshop does not seem to make a difference.

3. Let μ_1 = average number before coupon and μ_2 = average number when coupon is in effect; H_0: $\mu_1 = \mu_2$; H_1: $\mu_1 < \mu_2$; z_0 = -1.645; The sample proportion of plus signs r = 3/7 corresponds to z = -0.377. Since the sample test statistic does not fall in the critical region, we fail to reject H_0. There appears to be no difference in the average number of diners.

Chapter Test 12B

1. H_0: $\rho_s = 0$; H_1: $\rho_s > 0$; critical value $= 0.564$; Since the sample test statistic $r_s = 0.588$ falls in the critical region, we reject H_0. There seems to be a monotone increasing relation between midterm and final exam scores.

2. H_0: Income level makes no difference; H_1: Income level makes a difference; $\pm z_0 = \pm 1.96$; $\mu_R = 68$; $\sigma_R = 9.522$; R(less than \$30,000) $= 75$ corresponds to $z = 0.735$; Since the sample test statistic does not fall in the critical region, we fail to reject H_0. Income level does not seem to make a difference.

3. Let μ_1 = average before workshop and μ_2 = average after workshop; H_0: $\mu_1 = \mu_2$; H_1: $\mu_1 < \mu_2$; $z_0 = -1.645$; The sample proportion of plus signs $r = 0.375$ corresponds to $z = -0.707$; Since the sample test statistic does not fall in the critical region, we fail to reject H_0. The average confidence score seems to be the same.

Part III

Answers and Key Steps to Solutions of Even Numbered Problems

===============

Chapter 1

Section 1.1

2. Population: gender, working status, age of all students taking telecourses; sample: responses from 200 students surveyed
4. Population: shelf life of *all* Healthy Crunch granola bars; sample: shelf life of the 10 bars tested
6. Form B
8. (a) Ordinal (b) Ratio (c) Nominal
   (d) Interval (e) Ratio (f) Nominal
10. (a) Sampling (b) Simulation (c) Census
    (d) Experiment

Chapter Review Problems

2. Population: opinions of all listeners; sample: opinions of fifteen callers
4. Name, Social Security Number, color of hair, address, phone, place of birth, college major are all nominal; letter grade on test is ordinal; year of birth is interval; height and distance from home to college are ratio.

Chapter 2

Section 2.1

2. Answers vary; use groups of 3 digits.
4. Answers vary; use groups of 3 digits.
6. Answers vary; use groups of 4 digits.
8. (a) Assign each student a distinct ID number, and then use a random-number table.
   (b) Answers vary.
10. In all cases, assign distinct numbers to the items, and use a random-number table.
12. Assign participants distinct numbers from 01 to 62, and then use a random-number table to select 31 participants for group 1. The remaining participants will be in group 2.
14. Read 25 digits from the random-number table. When a digit is odd, record one outcome (heads), and when the digit is even, record the other outcome (tails).
16. Answers vary.
18. Read a sequence of 20 digits from the random-number table. Use true for even digits and false for odd digits.
20. (a) Stratified (b) Simple random (c) Cluster
    (d) Systematic (e) Convenience

2. Number of U.S. Bills Printed and Delivered in 1989 (in millions of bills)

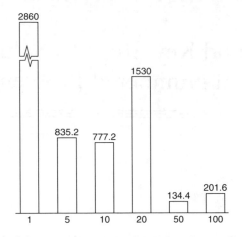

4. (a) Causes for Business Failure—Pareto Chart

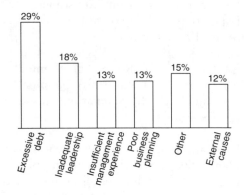

(b) No; debt

6. Meals We Are Most Likely to Eat in a Fast-Food Restaurant

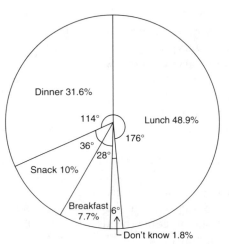

8. Distribution of the Tuition Dollar

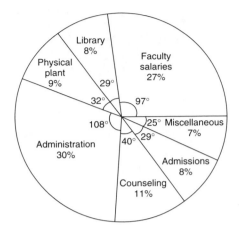

10. Driving Problems—Pareto Chart

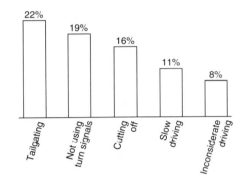

12. Social Security Tax as a Percentage of Wage, 1940–1990

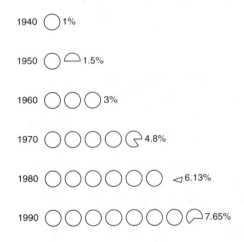

14. Amount of Money Spent on Medical Care in the United States (Billions of Dollars)

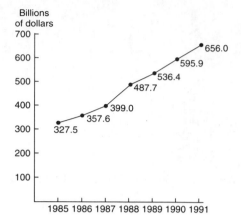

16. Changes in Boys' Height with Age

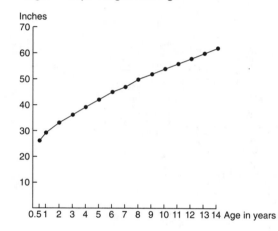

Section 2.3

2. (a) $cw = 10$

(b)

Percent Difficult Ski Terrain

Class Limits	Class Boundaries	Midpoint	Frequency	Relative Frequency	Cumulative Frequency
20–29	19.5–29.5	24.5	3	0.0857	3
30–39	29.5–39.5	34.5	6	0.1714	9
40–49	39.5–49.5	44.5	13	0.3714	22
50–59	49.5–59.5	54.5	9	0.2571	31
60–69	59.5–69.5	64.5	4	0.1144	45

(c & d) Percent Difficult Ski Terrain Histogram—Frequency Polygon

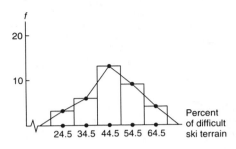

(e) Percent Difficult Ski Terrain—Relative-Frequency Histogram

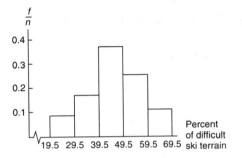

(f) Ogive for Percent Difficult Ski Terrain

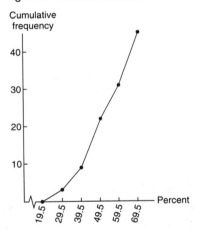

4. (a) $cw = 15$ (round up from 14 to 15)

(b) **Fast-Food Franchise Fees**

Class Limits	Class Boundaries	Midpoint	Frequency	Relative Frequency	Cumulative Frequency
5–19	4.5–19.5	12	21	0.3333	21
20–34	19.5–34.5	27	35	0.5555	56
35–49	34.5–49.5	42	5	0.0794	61
50–64	49.5–64.5	57	1	0.0159	62
65–79	64.5–79.5	72	1	0.0159	63

(c & d) Fees for Fast-Food Franchises—Histogram and Frequency Polygon (Thousands of Dollars)

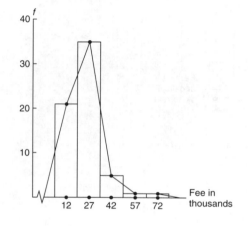

(e) Fees for Fast-Food Franchises—Relative-Frequency Histogram (Thousands of Dollars)

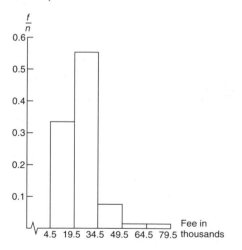

A–6

(f) Ogive for Fees

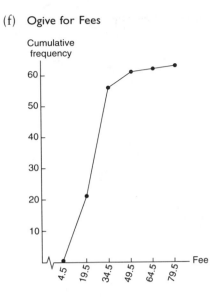

6. (a) $cw = 5$ (round up from 4 to 5)

(b) Length of Time on a Major League Team

Class Limits	Class Boundaries	Midpoint	Frequency	Relative Frequency	Cumulative Frequency
1–5	0.5–5.5	3	15	0.3261	15
6–10	5.5–10.5	8	16	0.3478	31
11–15	10.5–15.5	13	9	0.1957	40
16–20	15.5–20.5	18	5	0.1087	45
21–25	20.5–25.5	23	1	0.0217	46

(c & d) Length of Time on a Major League Team Histogram—Frequency Polygon

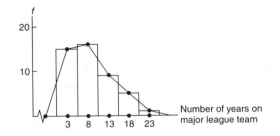

(e) Length of Time on a Major League Team—Relative-Frequency Histogram

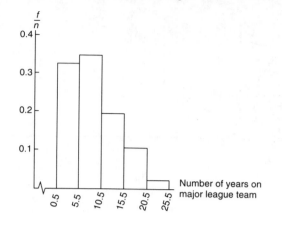

(f) Ogive for Time on Major League Team

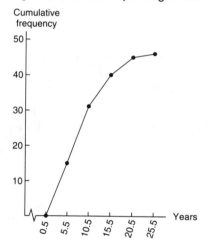

8. (a) $cw = 5$

(b) Commuters' Gasoline Usage

Class Limits	Class Boundaries	Midpoint	Frequency	Relative Frequency	Cumulative Frequency
3–7	2.5–7.5	5	6	0.1500	6
8–12	7.5–12.5	10	9	0.2250	15
13–17	12.5–17.5	15	12	0.3000	27
18–22	17.5–22.5	20	8	0.2000	35
23–27	22.5–27.5	25	4	0.1000	39
28–32	27.5–32.5	30	1	0.0250	40

(c & d) Commuters' Gasoline Usage—Histogram and Frequency Polygon

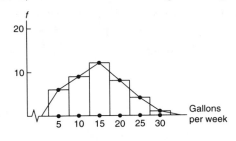

(e) Commuters' Gasoline Usage—Relative-Frequency Histogram

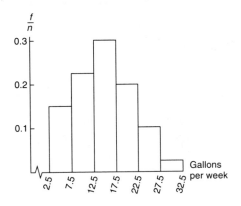

(f) Ogive for Gasoline Use

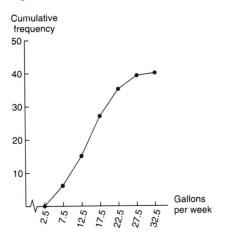

10. (a)   Miami Dolphins ($cw = 21$)

Class Limits	Midpoint	Frequency
175–195	185	13
196–216	206	7
217–237	227	19
238–258	248	8
259–279	269	11
280–300	290	12

San Diego Chargers ($cw = 32$)

Class Limits	Midpoint	Frequency
119–150	134.5	1
151–182	166.5	4
183–214	198.5	27
215–246	230.5	15
247–278	262.5	15
279–310	294.5	10

Weights of Football Players—Miami Dolphins

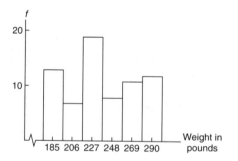

Weights of Football Players—San Diego Chargers

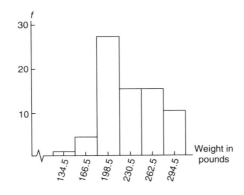

Weights of Football Players—San Diego Chargers

(b) Scales are different; use same limits.

12. (a) Ogive for Average Cost per Day

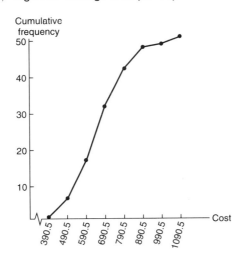

(b) 32
14. (a) Uniform; skewed right; bimodal; bimodal; symmetrical
    (b) Appeal to income group which occurs most frequently.
    (c) Somewhat questionable, since information is volunteered and cannot
         be checked.

10. (a)

Per Capita Income 1988

Unit = $100
19 | 1 = $19,000

```
11 | 1
12 | 2 2 3 3 5 7 7 8 8 8 9 9 9
13 | 6 9
14 | 3 6 7 7 8 9
15 | 0 1 3 3 5 5 5 8
16 | 2 5 5 6 6 7 8 9
17 | 5 6 7 7
18 | 9
19 | 1 3 4 5
20 | 8
21 |
22 | 0
23 | 1
```

Per Capita Income 1991

Unit = $100
19 | 1 = $19,100

```
11 |
12 |
13 | 3
14 | 2 5 8 8
15 | 1 1 4 4 5 6 8
16 | 0 1 3 4 4 6
17 | 1 3 4 5 5 6 7 8 9 9
18 | 0 5 7 8 9
19 | 1 1 2 4 4
20 | 0 3 8
21 | 0 0 3 9
22 | 1 5 9
23 |
24 |
25 | 4 9
```

(c) 1991 distribution shifts to higher incomes.

Chapter Review Problems

2. (a) about 322 to 332 ppm
   (b) about 345 to 355 ppm
   (c) about 23 ppm

4. (a)

Age of DUI Arrests

Unit = 1 year
1 | 6 = 16

```
1 | 6 8
2 | 0 1 1 2 2 2 3 4 4 5 6 6 6 7 7 7 9
3 | 0 0 1 1 2 3 4 4 5 5 6 7 8 9
4 | 0 0 1 3 5 6 7 7 9 9
5 | 1 3 5 6 8
6 | 3 4
```

(b) Class width = 7

Class Limits	Class Boundaries	Midpoint	Frequency	Relative Frequency	Cumulative Frequency
16–22	15.5–22.5	19	8	0.16	8
23–29	22.5–29.5	26	11	0.22	19
30–36	29.5–36.5	33	11	0.22	30
37–43	36.5–43.5	40	7	0.14	37
44–50	43.5–50.5	47	6	0.12	43
51–57	50.5–57.5	54	4	0.08	47
58–64	57.5–64.5	61	3	0.06	50

(c) Age Distribution of DUI Arrests Histogram

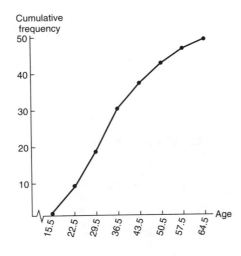

(d) Ogive for Age of DUI Arrests

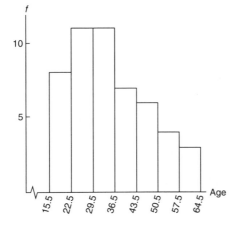

About 38% are 29 or under.

6. (a) Distribution of Civil Justice Caseloads Involving Business—Pareto Chart

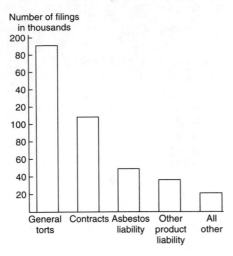

(b) Distribution of Civil Justice Caseloads Involving Business—Pie Chart

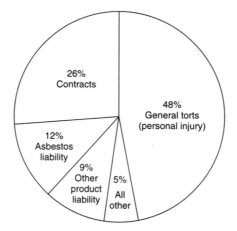

8. (a) Skewed left

(b)

Class Boundaries	Relative Frequency	Cumulative Frequency
0.75–1.25	1%	1%
1.25–1.75	1%	2%
1.75–2.25	2%	4%
2.25–2.75	8%	12%
2.75–3.25	17%	29%
3.25–3.75	27%	56%
3.75–4.25	44%	100%

(c) 29% less than 3.25; 56% less than 3.75

10. (a) Use a random-number table to obtain 30 distinct three-digit numbers between 001 and 950.

(b) Distribution of Makes of Cars in Parking Lot

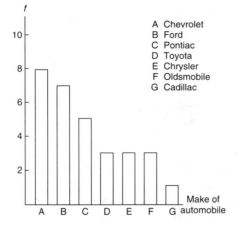

A Chevrolet
B Ford
C Pontiac
D Toyota
E Chrysler
F Oldsmobile
G Cadillac

Distribution of Makes of Cars in Parking Lot

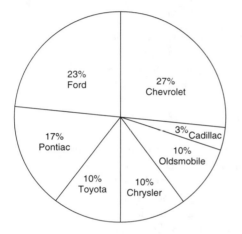

(c)

Class Limits	Midpoint	Frequency
70–75	72.5	1
76–81	78.5	4
82–87	84.5	13
88–93	90.5	10
94–99	96.5	2

Distribution of Model Years of Cars in Parking Lot

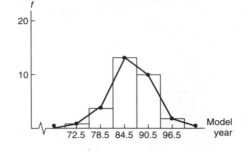

(d) In part b, nominal, cannot make histogram; interval data in part c; we can total the cars and make a circle graph showing the percentage in each model year interval.

12. (a) Cluster (b) Convenience (c) Systematic
    (d) Simple random (e) Stratified

Chapter 3

Section 3.1

2. Mean = 45.17; median = 46.5; mode = 46
4. Mean = 2.6; median = 2; mode = 1
6. (a) Mean = 11.0; median = 9.7; mode = 8
   (b) 9 ratings below the mean; 8 ratings below the median
   (c) OK, but try to do better.
   (d) Mean = 10.34; median = 9.3; a rating of 10 is better than half the shows got.
   (e) 10.7; higher, because a very low rating was dropped.
8. (a) Mean = 7.4; median = 8; mode = 8
   (b) Mean = 14.69; median = 8; mode = 8
   (c) Generally, the mean changes the most.
10. (a) Mean = 3.09; median = 2.9; mode = 3.3
    (b) Mean = 11.025; median = 6.75; mode = 6.6
    (c) The mean is very sensitive to extreme values.
12. (a) Mean, median, and mode if it exists
    (b) Mode if it exists
    (c) Mean, median, and mode if it exists
14. Discussion question
16. Answers may vary. Here are some examples
    (a) 1 2 2 2 3
    (b) 1 2 2 2 13
    (c) 1 1 5 5 5
    (d) 1 2 4 5 5

Section 3.2

2. (a) Range = 16.8; μ = 87.0; σ^2 = 38.9; σ = 6.2
   (b) CV = 7.1; the standard deviation is about 7.1% of the mean.
4. (a) Range = 37
   (b) $\bar{x}$ = 6.4
   (c) s = 12.1
6. (a) Range = 2.48; μ = 0.599; σ^2 = 0.561; σ = 0.749
   (b) CV = 125; the standard deviation is about one and one-quarter times as large as the mean.
8. (a) $\bar{x}$ = 109.5; s = 31.7; CV = 29.9; range = 69
   (b) $\bar{x}$ = 110.125; s = 7.2; CV = 6.5; range = 20
   (c) The first distribution is more spread than the second.
10. (a) Ralph: range = 4; Gloria: range = 7.2
    (b) Ralph: $\bar{x}$ = 21.6; s = 1.53; CV = 7.1; Gloria: $\bar{x}$ = 21.4; s = 3.22; CV = 15
    (c) Ralph got more consistent mileage, and his CV is lower.
12. (a) Results round to answers given.
    (b) 386 to 1074
    (c) 214 to 1246
14. Since CV = $s/\bar{x}$, then s = CV($\bar{x}$). s = (0.035)(4.8) = 0.168.

Section 3.3

2. (a) $\bar{x}$ ≈ 77.77   (b) s ≈ 1.45
4. (a) $\bar{x}$ ≈ 20.35   (b) s ≈ 3.703   (c) CV ≈ 18.2
6. (a) $\bar{x}$ ≈ 39.12   (b) s ≈ 17.01   (c) CV ≈ 43.48
8. (a) $\bar{x}$ ≈ 27; s ≈ 20.01; CV ≈ 74.1
   (b) $\bar{x}$ ≈ 18; s ≈ 13.29; CV ≈ 73.8
10. (a) $\bar{x}$ = 94.64   (b) s ≈ 118.8
12. (a) $\bar{x}$ ≈ 9.09   (b) s ≈ 3.30
14. (a) $\bar{x}$ ≈ 3.42   (b) s ≈ 1.90
16. 87.75; since the weights are the same, we could have computed the mean of the four scores directly.
18. (a) 8.09   (b) 8.18; this rating is higher.
20. (a) 8.53   (b) 9.38
    (c) Weighted average is sensitive to extreme values.
    (d) Answers vary.

Section 3.4

2. 75th percentile
4. Timothy
6. (a) About 50%   (b) About 90%   (c) About 40%
8. (a) Order the data.
   (b) Low = 1.6; Q_1 = 2.55; median = 3.0; Q_3 = 5.5; high = 7.0

(c) Percentage Increase in Faculty Salaries, Colorado—Box-and-Wisker Plot

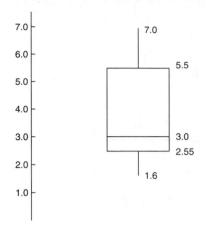

(d) There are several differences. An obvious one is that the middle half of the data for Colorado falls below that of Oregon.

10. (a) Order the data.

(b) Clerical Staff Length of Employment (Months)—Box-and-Wisker Plot

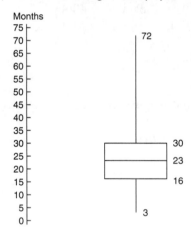

(c) IQR = 14

(d) Essay

12. (a) Order the data.

(b) Low = 280; Q_1 = 365; median = 450; Q_3 = 645; high = 830; IQR = 280

(c) Average Cost for Single-Lens Reflex Cameras (Dollars)

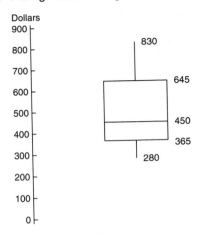

14. (a) Order the data.

(b) Number of Airline "No Shows" with Fares Paid in Advance

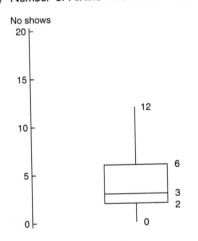

(c) IQR = 4
(d) Essay

16. (a) McDonald's   (b) Coca-Cola   (c) Coca-Cola
(d) Disney   (c) McDonald's   (d) Coca-Cola

18. (a) Order the data.
(b) Low = 11.5; Q_1 = 14.2; median = 15.25; Q_3 = 16.7; high = 22.0;
IQR = 2.5

(c) Average Disposable Income per Capita by State (Thousands of Dollars)

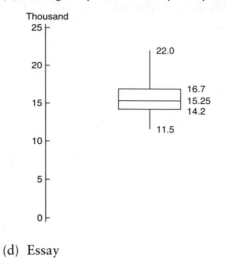

(d) Essay

Chapter Review Problems

2. (a) Medical Malpractice Claims (Millions of Dollars)

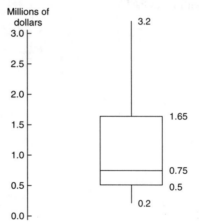

(b) $\bar{x} = 1.125$; median = 0.75; mean is larger.
(c) Range = 3; s = 0.90
4. (a) 85.77   (b) 82.17
6. (a) $\bar{x} = 4.53$; median = 4.05; mode = 1.9
   (b) s = 2.46; CV = 54.4; range = 6.7

8. (a) Speeds on the Atlanta Expressway

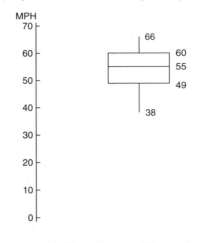

(b) $\bar{x} = 53.3$; median $= 55$; mode $= 60$
(c) True for the mean and median, but not for the mode.
10. $\bar{x} \approx 44.4$; $s \approx 11.8$
12. (a) $\mu = 437$; median $= 432$　　(b) No mode
 (c) Range $= 260$　　　　　　　　(d) $\sigma = 80.4$
14. (a) Zapper $\bar{x} = 0.94$; Zonker $\bar{x} = 1.14$
 (b) Zapper $s = 0.074$; Zonker $s = 0.19$
 (c) The Zapper had better average time with a smaller standard deviation, indicating that the Zapper performed more consistently.

Chapter 4

Section 4.1

2. Answers vary.
4. (a) Cannot be negative　　　　(b) Must be ≤ 1
 (c) $120\% = 1.20$ is too large.　　(d) Yes
6. (a) $P(0) = 0.04$; $P(1) = 0.19$; $P(2) = 0.33$; $P(3) = 0.35$; $P(4) = 0.09$
 (b) Yes (within round-off error)
8. (a) P(never) $= 0.20$; P(less than 1) $= 0.24$; P(1 to 2) $= 0.21$; P(more than 2) $= 0.35$
 (b) Yes (within round-off error)
10. (a) Yes, assuming no one wears both glasses and contact lenses
 (b) P(no glasses) $= 0.44$; P(no contacts) $= 0.964$
 (c) Neither $=$ complement of wears corrective lenses $= 0.404$
12. (a) 0.81　　(b) 0.19　　(c) Germinate or not germinate　　(d) No
14. (a) 0.46　　(b) 0.43　　(c) 0.20　　(d) 0.57

Section 4.2

2. (a) 0.5; yes   (b) 0.5; yes   (c) 0.6

4. (a) 0.05   (b) 0.66   (c) 0.29   (d) 0.35

6. (a) Yes   (b) 1/36   (c) 1/36   (d) 1/18

8. (a) 1/6   (b) 1/18   (c) Yes; 2/9

10. (a) No   (b) 0.006   (c) 0.006   (d) 0.012

12. (a) Yes   (b) 0.0059   (c) 0.0059   (d) 0.0118

14. (a) 0.0175   (b) 0.21   (c) 0.017; 0.012   (d) 0.07; 0.12

16. (a) 0.415 or 41.5%   (b) 0.962 or 96.2%

18. $P(\$600 \text{ or more}) = 0.283$; $P(\$199 \text{ or less}) = 0.371$

20. (a) 0.447; 0.319; 0.766   (b) 0.393; 0.380; 0.227
    (c) 0.388; 0.307; 0.304   (d) 0.084; 0.074; 0.100; 0.082
    (e) 0.766; 0.681; 0.553   (f) 0.211; 0.092; 0.033
    (g) No; 0.676; 0.484; 0.309

22. (a) 0.591; 0.466; 0.717   (b) No
    (c) 0.233; 0.359   (d) 0.409; 0.534
    (e) No   (f) 0.859

24. (a) 0.452; 0.333; 0.559   (b) 0.548; 0.667; 0.441
    (c) 0.472; 0.575; 0.348   (d) 0.528; 0.425; 0.652
    (e) 0.315; 0.685   (f) No
    (g) No

26. (a) 0.70 (b) 0.595 (c) 0.90 (d) 0.135 (e) 0.73
    (f) In the case of *and*, we are looking at the sample space of all students, both male and female. In the case of *given*, we are restricting our sample space to females only.

28. (a) 0.667   (b) 0.636; 0.727   (c) 0.424

30. (a) 0.27; 0.23; 0.73; 0.77   (b) 0.70; 0.95
    (c) 0.189; 0.69 (d) 0.11 (e) 0.69 (f) 0.189 (g) 0.31

Section 4.3

2. (a) Outcomes of Tossing a Coin and Throwing a Die

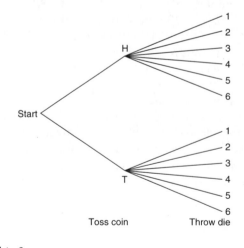

    (b) 2
    (c) 1/6

4. (a) Outcomes for Drawing Two Balls with Replacement

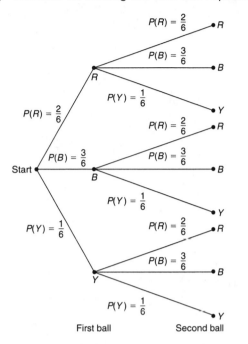

$P(R) = \frac{2}{6}$ •R

$P(B) = \frac{3}{6}$ •B

R

$P(Y) = \frac{1}{6}$ •Y

$P(R) = \frac{2}{6}$ •R

$P(R) = \frac{2}{6}$

$P(B) = \frac{3}{6}$ •B

$P(B) = \frac{3}{6}$

Start• B

$P(Y) = \frac{1}{6}$ •Y

$P(Y) = \frac{1}{6}$

$P(R) = \frac{2}{6}$ •R

$P(B) = \frac{3}{6}$ •B

Y

$P(Y) = \frac{1}{6}$ •Y

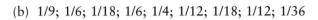

First ball   Second ball

(b) 1/9; 1/6; 1/18; 1/6; 1/4; 1/12; 1/18; 1/12; 1/36

6. (a) Outcomes of Three Multiple-Choice Questions

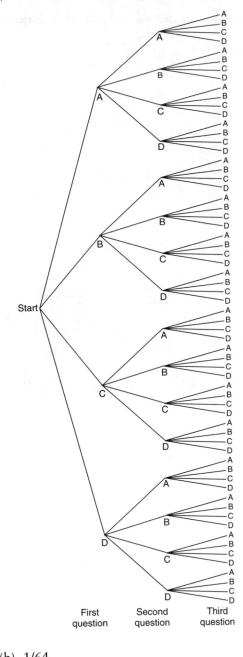

First
question

Second
question

Third
question

(b) 1/64

8. 24

10. (a) 36 (b) 9 (c) 0.25

12. 60

14. 336
16. 362,800
18. 56
20. 1
22. $P_{10,3} = 720$
24. $P_{6,6} = 720$
26. $C_{10,3} = 120$
28. (a) $C_{12,5} = 792$; (b) 0.001 (c) 0.027
30. (a) $C_{42,6} = 5,245,786$ (b) 0.00000019 (c) 0.0000019

Section 4.4

2. (a) Discrete   (b) Discrete   (c) Continuous
   (d) Continuous   (e) Continuous   (f) Discrete
4. (a) Discrete   (b) Continuous   (c) Continuous
   (d) Discrete   (e) Continuous   (f) Discrete

6. (a) Sizes of Families

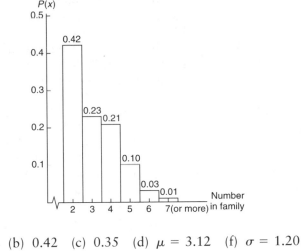

(b) 0.42   (c) 0.35   (d) $\mu = 3.12$ (f) $\sigma = 1.20$

8. (a) Number of Fish Caught in a 6-Hour Period at Pyramid Lake, Nevada

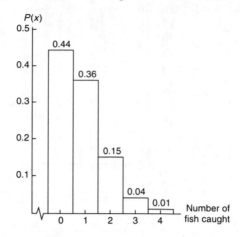

(b) 0.56   (c) 0.20   (d) $\mu = 0.82$ (f) $\sigma = 0.899$

10. (a) 0.746   (b) 0.346   (c) 0.013   (d) $\mu = 1.199$ (e) $\sigma = 0.9547$

12. (a) 0.008; 0.992
   (b) Expected earnings = $24, contribution = $21

14. (a) 0.015; 0.985
   (b) Expected earnings = $0.60, contribution = $3.60

Chapter Review Problems

2. (a) Sample space: 1H, 2H, 3H, 4H, 5H, 6H, 1T, 2T, 3T, 4T, 5T, 6T
   (b) Yes   (c) 0.167

4. (a) 0.125; 0.475; 0.3125; 0.0875   (b) 0.4
   (c) $\mu = 5.925$; $\sigma = 3.875$

6. (a) 0.470; 0.390; 0.140   (b) 0.840; 0.040
   (c) 0.100; 0.240   (d) 0.420; 0.060
   (e) 0.860; yes   (f) No

8. (a) 2, 3, 4, 5, 6, 7, 8, 9, 10, 11, 12

(b & c) Probability Distribution for the Sum of Two Dice

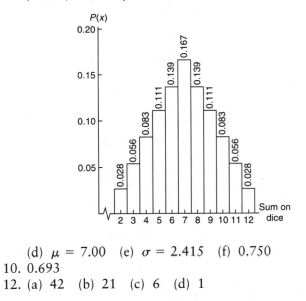

(d) $\mu = 7.00$ (e) $\sigma = 2.415$ (f) 0.750

10. 0.693
12. (a) 42   (b) 21   (c) 6   (d) 1

14. Ways to Satisfy Literature, Social Science, and Philosophy Requirements

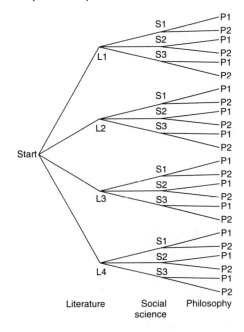

16. 64
18. $P_{3,3} = 6$

Chapter 5

2. Not binomial; the probability of success on one trial depends on which trip the client took.

4. Binomial   (a) A trial is giving a person medication.   (b) S = reduced blood pressure; F = no reduction in blood pressure.
   (c) $n = 100$; $p = 0.99$; $q = 0.01$; $r = 8$

6. Not binomial because the experiments were not repeated under identical conditions.

8. Binomial   (a) A trial is asking a single man if he would like a woman to take the initiative in asking for a date.   (b) S = response of yes; F = response of no.   (c) $n = 20$; $p = 0.71$; $q = 0.29$; $r = 18, 19, 20$

10. Not binomial; the probability of a drug dealing conviction might be different for local prisons, state prisons, and federal prisons. We are given information about federal prisons only.

12. Binomial   (a) A trial is asking an 18-year-old if he or she goes out on dates.   (b) S = response of yes; F = response of no.   (c) $n = 37$; $p = 0.79$; $q = 0.21$; $r = 29$ through 37.

14. Not binomial; the sample includes people who are not worth over 1 million dollars. If the sample were restricted to people worth over 1 million dollars, this would be a binomial experiment.

16. Not binomial; the probability of success on a trial depends on the weather.

2. $n = 10$; $p = 0.2$
   (a) 0.000 to three places after the decimal
   (b) 0.107
   (c) 0.892; 0.893; they should be equal, but because of round off error they differ slightly.

4. $n = 20$; $p = 0.7$
   (a) 0.001   (b) 0.417   (c) 0.048   (d) 0.745

6. $n = 20$; $p = 0.10$
   (a) 0.878   (b) 0.323   (c) 0.122   (d) 0.677

8. (a) $n = 9$; $p = 0.10$; 0.226; 0.387
   (b) $n = 4$; $p = 0.25$; 0.684; 0.004

10. (a) 0.683   (b) 0.633   (c) 0.166   (d) 0.001   (e) 0.003

12. (a) 0.002   (b) 0.410   (c) 0.028

14. (a) 0.663   (b) 0.460   (c) 0.043   (d) 0.001

16. (a) $P(6$ or more $M) = 0.740$; $P(6$ or more $F) = 0.473$;
    $P($less than 4 $F) = 0.135$
    (b) $P(6$ or more $M) = 0.961$; $P(6$ or more $F) = 0.117$;
    $P($less than 4 $F) = 0.493$

18. (a) 0.764   (b) 0.149   (c) 0.851   (d) 1 to three decimal places
    (e) 0.000 to three decimal places

20. (a) 0.000 to three decimal places   (b) 0.857   (c) 0.143

(d) 0.005   (e) 0.613   (f) 0.003

22. (a) 0.001524   (b) 0.0000119   (c) 0.0015359

Section 5.3

2. (a) II   (b) I   (c) III   (d) IV
   (e) More symmetrical when p is close to 0.5; skewed left when p is close to 1; skewed right when p is close to 0.

4. (a) Binomial Distribution for Number of Defective Syringes

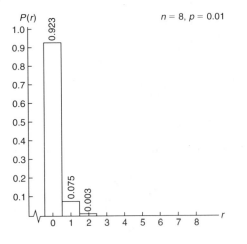

(b) $\mu = 0.08$ (c) 0.998   (d) $\sigma = 0.281$

6. (a) Binomial Distribution for Number of Automobile Damage Claims by People Under Age 25

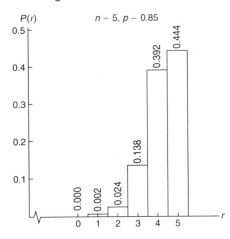

(b) $\mu = 4.25$; $\sigma = 0.798$; expected number is about 4.

8. (a) Binomial Distribution for Number of Automobiles Made by the "Big Three"

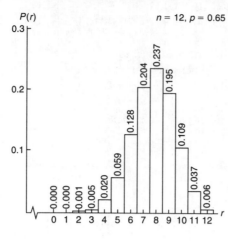

(b) $\mu = 7.8$; $\sigma = 1.65$ (c) 11   (d) 16

10. (a & b) Binomial Distribution for the Number of Hot Spots that are Forest Fires

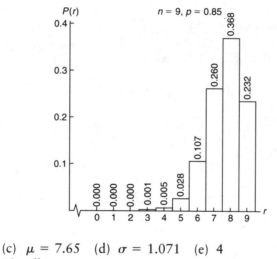

(c) $\mu = 7.65$ (d) $\sigma = 1.071$ (e) 4

12. 12 calls

14. (a) Vicky's Rifle Practice

(b) $\mu = 3.5$ (c) $\sigma = 1.025$ (e) $n = 10$

16. 15 bonds
18. 4 raincoats; for $n = 10$, $\mu = 6$
20. 9 inquiries; for $n = 25$, $\mu = 4.25$
22. (a) 0.008   (b) 0.028   (c) 0.836   (d) $\mu = 2.7$; $\sigma = 1.219$
    (e) 10 professors

Section 5.4

2. (a) $P(n) = (0.57)(0.43)^{n-1}$ (b) 0.2451   (c) 0.1054   (d) 0.0795
4. (a) $P(n) = (0.80)(0.20)^{n-1}$ (b) 0.8; 0.16; 0.032   (c) 0.008
   (d) 0.8847
6. (a) $P(n) = (0.036)(0.964)^{n-1}$ (b) 0.03345; 0.0311; 0.0241
   (c) 0.8636
8. (a) $\lambda = 7.5$ per 50 liter; $P(r) = e^{-7.5}(7.5)^r/r!$
   (b) 0.0156; 0.0389; 0.0729   (c) 0.9797   (d) 0.0203
10. (a) $\lambda = 3.7$ per 11 hours (rounded to nearest 10th)
    (b) 0.9753   (c) 0.7146   (d) 0.0247
12. (a) $\lambda = 7.2$ per 50 ft
    (b) 0.9745
    (c) $\lambda = 2.9$ per 20 ft; 0.5540
    (d) $\lambda = 0.3$ per 2 ft; 0.0037
    (e) Discussion
14. (a) Essay   (b) $\lambda = 1.00$ per 22 yr; 0.6321   (c) 0.3679
    (d) $\lambda = 2.27$ per 50 yr; 0.8967   (e) 0.1033
16. (a) $\lambda = 2.1$ (b) 0.1225   (c) 0.6203   (d) 0.6203
18. (a) $\lambda = 6.0$ (b) 0.9380; 0.5543
    (c) $\lambda = 13.0$; 0.0000 to four places; 0.9893; 0.6468
20. (a) $\lambda \approx 2.7$ (b) 0.0672   (c) 0.7513   (d) 0.2858

2. (a) 0.924   (b) 0.595
4. (a) 0.469   (b) 0.531

6. (a) Binomial Distribution of the Number of Engine Failures

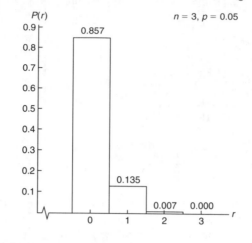

(b) 0.857; 0.007   (c) $\mu = 0.15$; $\sigma = 0.377$
8. 21.8 children
10. $\mu = 77.9$; $\sigma = 1.974$
12. $\mu = 255$
14. 0.996

16. (a) Binomial Distribution of the Number of Correct Answers on a True-False Test

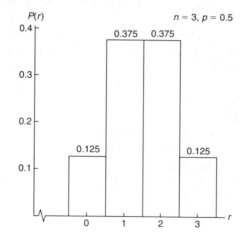

(b) $\mu = 1.5$ (c) $\sigma = 0.866$
18. (a) Essay   (b) $\lambda = 2.4$ per 100,000; 0.0907
    (c) $\lambda = 4.8$ per 200,000; 0.7058

Chapter 6

Section 6.1

2. $\mu = 16$; $\mu + \sigma = 18$; $\sigma = 2$

4. (a) Normal Curve

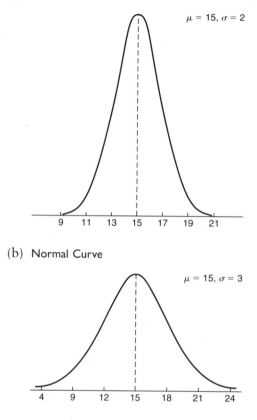

$\mu = 15, \sigma = 2$

(b) Normal Curve

$\mu = 15, \sigma = 3$

(c) Normal Curve

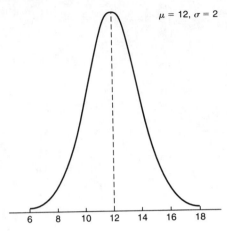

$\mu = 12, \sigma = 2$

(d) Normal Curve

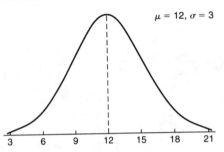

$\mu = 12, \sigma = 3$

    (e) No, the values of μ and σ are independent.

6. (a) 50% (b) 95.4% (c) 0.15%

8. (a) 95.4% or 954 chicks   (b) 68.2% or 682 chicks
   (c) 50% or 500 chicks   (d) 99.7% or 997 chicks

10. (a) 0.159 (b) 0.841 (c) About 136 cups

12. (a) 1.7 to 4.6 (b) 0.25 to 6.05

14. (a) Mean and standard deviation round to results shown.
   (b) 7.1% to 19.1% (c) 1.1% to 25.1% (d) Essay

16. (a) **Visitors Treated Each Day by YPMS (First Period)**

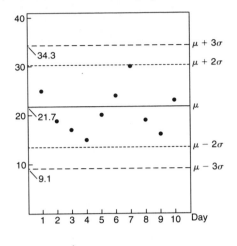

In control

(b) **Visitors Treated Each Day by YPMS (Second Period)**

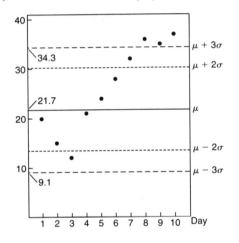

Out-of-control signals I and III are present.

18. (a) Number of Rooms Rented (First Period)

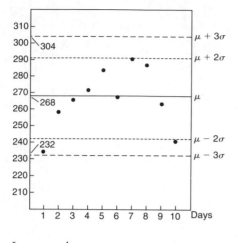

In control

(b) Number of Rooms Rented (Second Period)

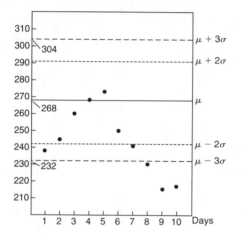

Out-of-control signals I and III are present.

20. Number of Breakfast Customers

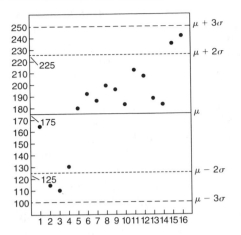

Out-of-control signals II and III are present.

Section 6.2

2. (a) Lee, Lang, Barnes (b) Smith, Anderson (c) Willis
   (d) Lee, 76; Willis, 73; Smith, 61; Lang, 96; Anderson, 66; Barnes, 92
4. (a) -0.67 (b) -2.17 (c) 0 (d) 2.17 (e) -2.50
   (f) 2.83
6. (a) $z < 0.65$ (b) $-1.91 < z$ (c) $1.12 < z < 1.81$
   (d) $17.9 < x$ (e) $x < 32.7$ (f) $18.6 < x < 33.4$
   (g) $z = -3.04$; this is very small.
   (h) $z = 3$ for very large fawn
8. (a) $100 \leqslant x \leqslant 115$ (b) $100 \leqslant x \leqslant 106.3$ (c) $69.4 \leqslant x \leqslant 131.1$
   (d) $70.6 \leqslant x$ (e) $x \leqslant 138.7$ (f) $x \geqslant 114.1$
10. (a) $0.86 < z$ (b) $z < -0.86$ (c) $-2.29 < z < -1.71$
    (d) $x < 9513$ (e) $11{,}332 < x$ (f) $7938 < x < 9688$
    (g) Unusually low, since $z = -2.9$.
12. (a) Site 1, $z = -0.63$; site 2, $z = 2.80$
    (b) Site 2 is more unusual.
14. Mary, $z = 1.70$; Harold, $z = 0.65$; Mary did better.

Section 6.3

2. 0.4982	4. 0.4732
6. 0.8980	8. 0.0628
10. 0.4641	12. 0.5000
14. 0.4404	16. 0.3192
18. 0.9850	20. 0.7642
22. 0.2054	24. 0.4911
26. 0.0409	28. 0.0718
30. 0.8369	32. 0.5000
34. 0.0150	36. 0.0158
38. 0.9332	40. 0.9993

Section 6.4

2. 0.8914   4. 0.0471
6. 0.1693   8. 0.0918
10. ~0.9999   12. 1.96
14. −0.95   16. −1.63
18. 1.645   20. ±1.96
22. (a) 0.9664 (b) 0.9664 (c) 0.9328 (d) 0.0336
24. (a) SAT of 628; ACT of 26
    (b) SAT of 584; ACT of about 23
    (c) SAT of 475; ACT of 16.5
26. (a) 0.8036 (b) 0.0228 (c) 0.1736
28. (a) 0.0228 (b) 0.2420 (c) 0.2061
30. (a) About 21.2% (b) 22 months
32. (a) $\sigma \approx 1.5$ (b) 0.9772 (c) 0.9082 (d) About 6.1 years
34. (a) $\sigma \approx 1$ (b) 0.0001 (c) 0.8413 (d) About 4 years
36. (a) 61.8 hours (b) 58.8 hours (c) 53.4 hours

Chapter Review Problems

2. (a) 0.2734 (b) 0.4332 (c) 0.0371 (d) 0.0594
   (e) 0.1660 (f) 0.9778
4. (a) 0.7967 (b) 0.9938 (c) 0.2865
6. −2.33
8. ±2.58 (Note that we do not interpolate because the areas are changing very slowly this far into the tails of the curve.)
10. (a) 336.5 (b) 261.25 (c) 0.9544
12. (a) 0.5000 (b) 4260 hours

14. (a) Hydraulic Pressure in Main Cylinder of Landing Gear of Airplanes (psi)— First Data Set

In control

(b) Hydraulic Pressure in Main Cylinder of Landing Gear of Airplanes (psi)—
Second Data Set

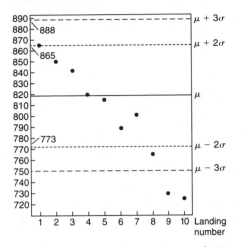

Out of control with signals I and III present

16. (a) 0.6103   (b) 0.9573   (c) 0.6872
18. 36 months to the nearest month so that no more than 3% need to be replaced.

Chapter 7

Section 7.1

2. See Sec. 2.1.
4. A numerical descriptive measure of a sample. Examples: $\bar{x}$, s, s^2, $\hat{p}$, and so forth.
6. A probability distribution for a sample statistic.
8. Relative frequencies can be thought of as a measure or estimate of the likelihood of a certain statistic falling within the class bounds.
10. We studied the sampling distribution of mean trout lengths based on samples of size 5. Other such sampling distributions abound.

Section 7.2

Note: Answers may differ slightly depending on how many digits are carried in the standard deviation.

2. (a) Mean = 100; standard deviation = 5.33; 0.4332
   (b) Mean = 100; standard deviation = 4.36; 0.4664
   (c) The standard deviation is smaller for part b.
4. (a) No, sample size is too small.
   (b) Normal with mean 72 and standard deviation 2; 0.6687
6. (a) 0.0793   (b) 0.0174   (c) 0.9033
8. (a) 0.9948   (b) 0.9345   (c) 0.9293
10. (a) Miami 0.7019; Fairbanks 0.7157
    (b) Miami 0.9177; Fairbanks 0.9322

(c) Miami 0.9983; Fairbanks 0.9992
12. (a) 0.0110; 0.0228   (b) 0.0006; 0.0023   (c) Less than 0.0001; 0.0003
    (d) The probabilities decreased as n increased. It would be an extremely rare event for a person to have two or three tests below 3500 purely by chance.
14. (a) Normal with mean 16 and standard deviation 0.3651.
    (b) 0.4969   (c) 0.0031
16. (a) 0.2586   (b) 0.6826
    (c) The standard deviation is smaller for the $\bar{x}$ distribution.
18. (a) 0.8238   (b) 0.9808   (c) 0.8046
20. (a) 0.8212; 0.0154   (b) 0.9993; no
22. (a) Approximately normal with mean $20 and standard deviation $0.7.
    (b) 0.9958   (c) 0.2282   (d) Yes

Section 7.3

Note: Answers may differ slightly depending on how many digits are carried in the computation of the standard deviation and the computation of z.

2. (a) greater than 0.9999   (b) 0.9798   (c) 0.8594
4. (a) 0.8023   (b) 0.9671   (c) 0.8156   (d) Yes
6. (a) 0.7881   (b) 0.9936   (c) 0.7817   (d) Yes
8. (a) 0.9370   (b) 0.9842   (c) 0.9633   (d) 0.7157
10. (a) 0.0571   (b) 0.0021   (c) 0.9408   (d) Yes
12. (a) 0.0274   (b) 0.8739   (c) 0.1251
14. (a) 0.0708   (b) 0.0708   (c) 0.9972
16. (a) Less than 0.0001   (b) Less than 0.0001   (c) 0.8508

Chapter Review Problems

2. All the $\bar{x}$ distributions will be normal with mean 15. The standard deviations will be 3/2, 3/4, and 3/10, respectively.
4. (a) Both np and nq are greater than 5.
   (b) 0.6678; 0.668; they agree to three decimal places.
6. (a) 0.2743   (b) 0.0287
   (c) The standard deviation of $\bar{x}$ is smaller.
8. (a) 0.9082   (b) 0.9332
   (c) The standard deviation of part b is smaller.
10. (a) 0.0268   (b) 0.8384
12. (a) 0.8643   (b) 0.7286
14. 0.8664
16. (a) 0.2676   (b) 0.0110   (c) 0.7214

Chapter 8

Section 8.1

2. (a) 14.05 to 17.37   (b) 14.63 to 16.53
   (c) 15.07 to 16.11   (d) Lengths decrease as n increases.
4. (a) Calculator results round to given answers. The confidence intervals are computed using the rounded results.
   (b) 20.85 to 26.07   (c) Yes
   (d) 20.11 to 26.81; yes
6. (a) 9.20 to 10.28   (b) 1058 to 1182
8. (a) 92.5 to 101.5   (b) Go up.
10. (a) 5.43 to 5.67   (b) 1.97 to 2.09
12. (a) 18.99 to 20.01   (b) 22.04 to 23.56
14. (a) 8.46 to 8.94; 8.39 to 9.01; 8.33 to 9.07; 8.21 to 9.19
    (b) As c increases, the lengths of the intervals also increase.
16. 9.87 to 11.13
18. (a) 46.49 to 52.09
    (b) In this community, the cost is higher.

Section 8.2

2. 5.841
4. 2.201
6. (a) Use calculator.   (b) 2.59 to 4.57
8. (a) Use calculator.   (b) 18.93 to 20.55   (c) Above average
10. (a) Use calculator.   (b) 5.00 to 7.90
12. $\bar{x} = 244.5$; $s = 21.73$; 217.6 to 271.4
14. 12.6 to 13.6
16. (a) 3.48 to 6.34   (b) 5.05 to 6.43
    (c) Second interval is shorter because the s is smaller and n is larger.
18. (a) Use calculator.   (b) 18.6 to 22.6

Section 8.3

2. (a) $\hat{p} = 0.5491$ (b) 0.49 to 0.61   (c) Yes
4. (a) $\hat{p} = 0.6081$ (b) 0.57 to 0.65   (c) Yes
6. (a) $\hat{p} = 0.1600$ (b) 0.12 to 0.20   (c) Yes
8. (a) $\hat{p} = 0.7600$ (b) 0.75 to 0.77   (c) Yes
10. (a) $\hat{p} = 0.6400$ (b) 0.62 to 0.66   (c) Yes
12. (a) $\hat{p} = 0.8201$ (b) 0.80 to 0.84
14. (a) $\hat{p} = 0.2500$ (b) 0.23 to 0.27
16. (a) $\hat{p} = 0.3177$ (b) 0.28 to 0.35
18. (a) $\hat{p} = 0.1100$ (b) 0.07 to 0.15
20. As part of its 5-year literacy campaign, Coors Brewing Company commissioned a poll of 1000 adults. About 25% of them read at least 30 minutes before retiring each night. The margin of error is about 2.7 percentage points.

Section 8.4

2. 866
4. 76 total or 35 more

6. (a) 456   (b) 257
8. (a) 106   (b) 66
10. 349 total or 199 more
12. 68 total or 28 more
14. 326 total or 291 more
16. (a) 97   (b) 52 total or 14 more
18. (a) 271   (b) 214
20. 1068

Section 8.5

2. (a) Use calculator.   (b) $s = 41.67$; 31.74 to 88.92
   (c) Average print shop startup costs greater.
4. (a) $s = 3.36$; -1.3 to 2.34   (b) $s = 3.42$; 0.11 to 3.81
   (c) $s = 3.68$; -0.55 to 3.43
   (d) No difference between competence and social acceptance. Competence seems to exert more influence than attractiveness. No difference between social acceptance and attractiveness.
6. (a) $\hat{p}_1 = 0.3520$; $\hat{p}_2 = 0.3800$; $\hat{\sigma} = 0.0320$; -0.08 to 0.02
   (b) No difference
8. (a) Use calculator.   (b) $s = 8.6836$; 3.89 to 14.05
   (c) Average weight of gray wolves from Chihuahua is greater than those in Durango.
10. (a) $\hat{p}_1 = 0.3095$; $\hat{p}_2 = 0.1184$; $\hat{\sigma} = 0.0413$; 0.08 to 0.30
    (b) Greater proportion of hogans occurs at Fort Defiance.
12. (a) $\hat{p}_1 = 0.6161$; $\hat{p}_2 = 0.1857$; $\hat{\sigma} = 0.05650$; 0.28 to 0.58
    (b) Greater proportion of artifacts seems to be unidentified at higher elevations.
14. (a) $\hat{p}_1 = 0.8196$; $\hat{p}_2 = 0.2243$; $\hat{\sigma} = 0.0297$; 0.57 to 0.63
    (b) It seems that the treatment does make a difference.
16. (a) $\hat{p}_1 = 0.5696$; 0.53 to 0.61   (b) $\hat{p}_2 = 0.3354$; 0.30 to 0.37
    (c) $\hat{\sigma} = 0.0282$; 0.18 to 0.29; separated nesting boxes yield more wood ducks.

Chapter Review Problems

2. Mean, large sample; 729 to 770
4. Sample size for mean; 102
6. Mean; small sample; 12.8 to 14.2
8. Sample size for proportion; 9589
10. Mean, large sample; 51.1 to 55.1
12. Proportion; $\hat{p} = 0.2333$; 0.18 to 0.28
14. Mean, small sample; 3.55 to 3.65
16. Proportion: $\hat{p} = 0.70$; 0.69 to 0.71
18. Mean, large sample; 38.2 to 47.8
20. Proportion; $\hat{p} = 0.40$; 0.32 to 0.48
22. Mean, small sample; 206.5 to 239.5
24. Sample size for means; 167 total or 67 more
26. mean, small sample; 44711 to 68763

28. (a) Difference of means, large samples; -1.41 to -0.71
  (b) The 60-second ads seem to receive a higher average rating.
30. Difference of means, small samples
  (a) Use calculator.   (b) $s = 3.3595$; -3.6 to 1.8; no difference
32. Difference of proportions
  (a) $\hat{p}_1 = 0.8495$; $\hat{p}_2 = 0.8916$; -0.14 to 0.06 (b) No difference

Chapter 9

Section 9.1

2. Alternate hypothesis
4. No
6. (a) H_0: $\mu = 2.7$
  (b) H_1: $\mu > 2.7$; right of the mean
  (c) 0.05; type I
  (d) Critical value, critical region, and sample test statistic
8. (a) H_0: $\mu = 16.4$ (b) H_1: $\mu > 16.4$ (c) H_1: $\mu < 16.4$ (d) H_1: $\mu \neq$ 16.4
  (e) Part b to right of mean; part c to left of mean; part d on both sides of mean
  (f) Critical value, critical region, sample test statistic, level of significance
10. (a) H_0: $\mu = 288$ (b) H_1: $\mu > 288$, right-tailed   (c) H_1: $\mu < 288$, left-tailed
  (d) H_1: $\mu \neq 288$, both tails
  (e) Significance level, critical value(s), critical region, sample test statistic

Section 9.2

2. H_0: $\mu = 27{,}600$; H_1: $\mu > 27{,}600$; $z_0 = 1.645$; $z = 1.38$; fail to reject H_0
4. H_0: $\mu = 3218$; H_1: $\mu > 3218$; $z_0 = 2.33$; $z = 3.93$; reject H_0
6. H_0: $\mu = 1323$; H_1: $\mu \neq 1323$; $\pm z_0 = \pm 2.58$; $z = -2.29$; fail to reject H_0
8. H_0: $\mu = 159$; H_1: $\mu < 159$; $z_0 = -2.33$; $z = -3.14$; reject H_0
10. H_0: $\mu = 23.6$; H_1: $\mu > 23.6$; $z_0 = 1.645$; $z = 3.67$; reject H_0
12. H_0: $\mu = 28.76$; H_1: $\mu > 28.76$; $z_0 = 2.33$; $z = 3.20$; reject H_0
14. H_0: $\mu = 1.42$; H_1: $\mu > 1.42$; $z_0 = 1.645$; $z = 1.48$; fail to reject H_0
16. H_0: $\mu = 61{,}400$; H_1: $\mu < 61{,}400$; $z_0 = -1.645$; $z = -1.92$; reject H_0
18. (a) One-tailed test   (b) Two-tailed test   (c) Yes
20. (a) 20.28 to 23.72; since the hypothesized value of μ, 21, lies in the confidence interval, do not reject H_0.
  (b) $\pm z_0 = \pm 2.58$; $z = 1.5$; do not reject H_0.

Section 9.3

2. (a) Fail to reject H_0 (b) Fail to reject H_0
4. (a) Fail to reject H_0 (b) Reject H_0
6. (a) Fail to reject H_0 (b) Fail to reject H_0
8. $z = 1.87$; P value $= 0.0307$; reject H_0 at 5% level; fail to reject H_0 at 1% level
10. $z = -0.57$; P value $= 0.2843$; fail to reject H_0 at both 1% and 5% levels
12. $z = 0.44$; P value $= 0.66$; fail to reject H_0 at both the 1% and 5% levels
14. $z = 2.53$; P value $= 0.0057$; reject H_0 at both the 1% and 5% levels

Section 9.4

2. 2.681
4. -1.740
6. 2.467
8. (a) Use calculator.
   (b) H_0: $\mu = 7500$; H_1: $\mu \neq 7500$; $\pm t_0 = \pm 2.365$; $t = -4.32$; P value < 0.010; reject H_0
10. (a) Use calculator.
    (b) H_0: $\mu = 14$; H_1: $\mu > 14$; $t_0 = 2.718$; $t = 5.535$; P value < 0.005; reject H_0
12. (a) Use calculator.
    (b) H_0: $\mu = 3.2$; H_1: $\mu \neq 3.2$; $\pm t_0 = \pm 2.262$; $t = 1.456$; $0.15 < P$ value < 0.20; fail to reject H_0
14. (a) Use calculator.
    (b) H_0: $\mu = 687$; H_1: $\mu > 687$; $t_0 = 1.796$; $t = 2.095$; $0.025 < P$ value < 0.05; reject H_0
16. (a) Use calculator.
    (b) H_0: $\mu = 600$; H_1: $\mu < 600$; $t_0 = -1.383$; $t = -1.595$; $0.050 < P$ value < 0.075; reject H_0 at 10% level of significance
18. (a) Use calculator.
    (b) H_0: $\mu = 6.3$; H_1: $\mu > 6.3$; $t_0 = 2.132$; $t = 0.787$; $0.125 < P$ value; fail to reject H_0

Section 9.5

2. H_0: $p = 0.221$; H_1: $p < 0.221$; $z_0 = -1.645$; $z = -1.85$ for $\hat{p} = 32/193$; P value $= 0.0322$; reject H_0
4. H_0: $p = 0.39$; H_1: $p \neq 0.39$; $\pm z_0 = \pm 1.96$; $z = 0.50$ for $\hat{p} = 128/317$; P value $= 0.617$; fail to reject H_0
6. H_0: $p = 0.67$; H_1: $p > 0.67$; $z_0 = 1.645$; $z = 1.90$ for $\hat{p} = 42/53$; P value $= 0.0287$; reject H_0
8. H_0: $p = 0.75$; H_1: $p \neq 0.75$; $\pm z_0 = \pm 1.96$; $z = 1.50$ for $\hat{p} = 94/116$; P value $= 0.1336$; fail to reject H_0
10. H_0: $p = 0.75$; H_1: $p \neq 0.75$; $\pm z_0 = \pm 1.96$; $z = 0.44$ for $\hat{p} = 64/83$; P value $= 0.660$; fail to reject H_0
12. H_0: $p = 0.47$; H_1: $p \neq 0.47$; $\pm z_0 = \pm 2.58$; $z = -2.80$ for $\hat{p} = 81/216$; P value $= 0.0052$; reject H_0
14. H_0: $p = 0.80$; H_1: $p < 0.80$; $z_0 = -1.645$; $z = -0.93$ for $\hat{p} = 88/115$; P value $= 0.1762$; fail to reject H_0
16. H_0: $p = 0.24$; H_1: $p > 0.24$; $z_0 = 2.33$; $z = 2.06$ for $\hat{p} = 23/66$; P value $= 0.0197$; fail to reject H_0
18. H_0: $p = 0.863$; H_1: $p \neq 0.863$; $\pm z_0 = \pm 1.96$; $z = 2.17$ for $\hat{p} = 891/1005$; P value $= 0.03$; reject H_0

Section 9.6

2. $\bar{d} = 2.26$; $s_d = 2.204$; H_0: $\mu_d = 0$; H_1: $\mu_d > 0$; $t_0 = 2.132$; $t = 2.293$; $0.025 < P$ value < 0.05; reject H_0
4. $\bar{d} = 0.08$; $s_d = 1.701$; H_0: $\mu_d = 0$; H_1: $\mu_d > 0$; $t_0 = 2.812$; $t = 0.1487$; $0.125 < P$ value; fail to reject H_0

6. $\bar{d} = -0.84$; $s_d = 3.57$; H_0: $\mu_d = 0$; H_1: $\mu_d \neq 0$; $\pm t_0 = \pm 3.106$; $t = -0.815$; $0.25 < P$ value; fail to reject H_0

8. $\bar{d} = 0.0$; $s_d = 8.76$; H_0: $\mu_d = 0$; H_1: $\mu_d > 0$; $t_0 = 1.943$; $t = 0.000$; $0.125 < P$ value; fail to reject H_0

10. $\bar{d} = 1.25$; $s_d = 1.91$; H_0: $\mu_d = 0$; H_1: $\mu_d > 0$; $t_0 = 1.895$; $t = 1.851$; $0.05 < P$ value < 0.075; fail to reject H_0

12. $\bar{d} = 6.33$; $s_d = 7.174$; H_0: $\mu_d = 0$; H_1: $\mu_d > 0$; $t_0 = 2.015$; $t = 2.161$; $0.025 < P$ value < 0.05; reject H_0

14. $\bar{d} = 0.4$; $s_d = 0.447$; H_0: $\mu_d = 0$; H_1: $\mu_d > 0$; $t_0 = 2.015$; $t = 2.192$; $0.025 < P$ value < 0.05; reject H_0

16. $\bar{d} = 0.1111$ $s_d = 5.2784$; H_0: $\mu_d = 0$; H_1: $\mu_d \neq 0$; $\pm t_0 = \pm 2.306$; $t = 0.0631$; $0.25 < P$ value; fail to reject H_0

18. $\bar{d} = -3.33$; $s_d = 7.3394$; H_0: $\mu_d = 0$; H_1: $\mu_d \neq 0$; $\pm t_0 = \pm 2.571$; $t = -1.111$; $0.250 < P$ value; fail to reject H_0

Section 9.7

2. H_0: $\mu_1 = \mu_2$; H_1: $\mu_1 \neq \mu_2$; $\pm z_0 = \pm 2.58$; $z = -2.127$ for $\bar{x}_1 - \bar{x}_2 = -8$; P value $= 0.0332$; fail to reject H_0

4. H_0: $\mu_1 = \mu_2$; H_1: $\mu_1 \neq \mu_2$; $\pm z_0 = \pm 1.96$; $z = 1.67$ for $\bar{x}_1 - \bar{x}_2 = 8$; P value $= 0.0950$; fail to reject H_0

6. H_0: $\mu_1 = \mu_2$; H_1: $\mu_1 \neq \mu_2$; $\pm z_0 = \pm 2.58$; $z = -0.86$ for $\bar{x}_1 - \bar{x}_2 = -0.4$; P value $= 0.3898$; fail to reject H_0

8. H_0: $\mu_1 = \mu_2$; H_1: $\mu_1 > \mu_2$; $z_0 = 2.33$; $z = 1.52$ for $\bar{x}_1 - \bar{x}_2 = 19.2$; P value $= 0.0643$; fail to reject H_0

10. (a) Use calculator.
    (b) H_0: $\mu_1 = \mu_2$; H_1: $\mu_1 \neq \mu_2$; $\pm t_0 = \pm 2.110$; $s = 139.323$; $t = -0.828$ for $\bar{x}_1 - \bar{x}_2 = -53$; $0.25 < P$ value; fail to reject H_0

12. (a) Use calculator.
    (b) H_0: $\mu_1 = \mu_2$; H_1: $\mu_1 > \mu_2$; $t_0 = 2.518$; $s = 82.902$; $t = 0.991$ for $\bar{x}_1 - \bar{x}_2 = 31.51$; $0.125 < P$ value; fail to reject H_0

14. H_0: $\mu_1 = \mu_2$; H_1: $\mu_1 < \mu_2$; $t_0 = -2.492$; $s = 7.895$; $t = -1.914$ for $\bar{x}_1 - \bar{x}_2 = -6$; $0.025 < P$ value < 0.05; fail to reject H_0

16. (a) Use calculator.
    (b) H_0: $\mu_1 = \mu_2$; H_1: $\mu_1 < \mu_2$; $t_0 = -1.812$; $s - 2.8$; $t = -1.732$ for $\bar{x}_1 - \bar{x}_2 = -2.8$; $0.05 < P$ value < 0.075; fail to reject H_0

18. $\hat{p}_1 = 0.1301$; $\hat{p}_2 = 0.0602$; H_0: $p_1 = p_2$; H_1: $p_1 > p_2$; $z_0 = 1.645$; $z = 5.67$; P value < 0.0001; reject H_0

20. $\hat{p}_1 = 0.7800$; $\hat{p}_2 = 0.2000$; H_0: $p_1 = p_2$; H_1: $p_1 > p_2$; $z_0 = 2.33$; $z = 8.204$; P value < 0.0001; reject H_0

22. $\hat{p}_1 = 0.4709$; $\hat{p}_2 = 0.6202$; H_0: $p_1 = p_2$; H_1: $p_1 < p_2$; $z_0 = -1.645$; $z = -4.44$; P value < 0.0001; reject H_0

24. $\hat{p}_1 = 0.1636$; $\hat{p}_2 = 0.1776$; H_0: $p_1 = p_2$; H_1: $p_1 \neq p_2$; $\pm z_0 = \pm 2.58$; $z = -0.59$; P value $= 0.5552$; fail to reject H_0

Chapter Review Problems

2. Single proportion; H_0: $p = 0.25$; H_1: $p \neq 0.25$; $\pm z_0 = \pm 2.58$; $z = 1.80$ for $\hat{p} = 18/50$; P value $= 0.0718$; fail to reject H_0

4. Single proportion; H_0: $p = 0.35$; H_1: $p > 0.35$; $z_0 = 1.645$; $z = 2.48$ for $\hat{p} = 39/81$; P value $= 0.0066$; reject H_0

6. Difference of proportions; $\hat{p}_1 = 0.5645$; $\hat{p}_2 = 0.6613$; H_0: $p_1 = p_2$; H_1: $p_1 < p_2$; $z_0 = -2.33$; $z = -1.11$; P value = 0.1335; fail to reject H_0

8. Single proportion; H_0: $p = 0.60$; H_1: $p < 0.60$; $z_0 = -2.33$; $z = -3.10$ for $\hat{p} = 40/90$; P value = 0.0010; reject H_0

10. Single mean; H_0: $\mu = 70$; H_1: $\mu > 70$; $z_0 = 2.33$; $z = 1.29$; P value = 0.0985; fail to reject H_0

12. Difference of means, large samples; H_0: $\mu_1 = \mu_2$; H_1: $\mu_1 \neq \mu_2$; $\pm z_0 = \pm 1.96$; $z = -3.48$ for $\bar{x}_1 - \bar{x}_2 = -9$; P value = 0.0003; reject H_0

14. Difference of proportions; $\hat{p}_1 = 0.1364$; $\hat{p}_2 = 0.1856$; H_0: $p_1 = p_2$; H_1: $p_1 < p_2$; $z_0 = -1.645$; $z = -0.91$; P value = 0.1814; fail to reject H_0

16. Single proportion; H_0: $p = 0.51$; H_1: $p < 0.51$; $z_0 = -1.645$; $z = -2.26$ for $\hat{p} = 86/200$; P value = 0.0119; reject H_0

18. Single proportion; H_0: $p = 0.36$; H_1: $p < 0.36$; $z_0 = -1.645$; $z = -1.94$ for $\hat{p} = 33/120$; P value = 0.0262; reject H_0

20. Difference of proportions; $\hat{p}_1 = 0.0612$; $\hat{p}_2 = 0.0357$; H_0: $p_1 = p_2$; H_1: $p_1 > p_2$; $z_0 = 2.33$; $z = 1.18$; P value = 0.1190; fail to reject H_0

22. Paired difference test; $\bar{d} = 4.94$; $s_d = 3.901$; H_0: $\mu_d = 0$; H_1: $\mu_d > 0$; $t_0 = 2.132$; $t = 2.832$; $0.01 <$ P value < 0.025; reject H_0

24. Difference of means, small samples; H_0: $\mu_1 = \mu_2$; H_1: $\mu_1 \neq \mu_2$; $\pm t_0 = \pm 2.101$; $s = 6.4950$; $t = 0.2066$ for $\bar{x}_1 - \bar{x}_2 = 0.6$; $0.25 <$ P value; fail to reject H_0

26. (a) Reject H_0 (b) Reject H_0

Chapter 10

Section 10.1

2. No linear correlation
4. Moderate linear correlation
6. No linear correlation

8. (a) List Price and Best Price for Models of Chevrolet Cavalier (Thousands of Dollars)

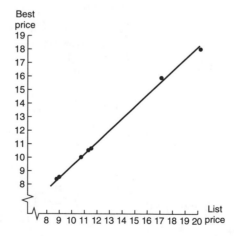

(b) Draw the line that you think fits best (specific equation in Sec. 10.2).

(c) High

10. (a) Group Health Insurance Plans: Average Number of Employees versus Administrative Costs as a Percentage of Claims

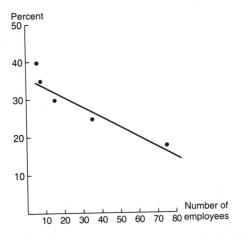

(b) Draw the line that you think fits best (specific equation in Sec. 10.2).

(c) Moderate

12. (a) Incidence of Malignant Melanoma after Peaks of Sunspot Activity

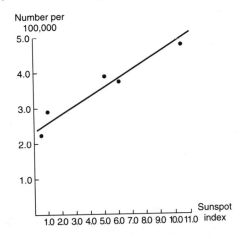

(b) Draw the line that you think fits best (specific equation in Sec. 10.2).

(c) Moderate

14. (a)   Distance on 1 Gallon of Gasoline at Different Speeds

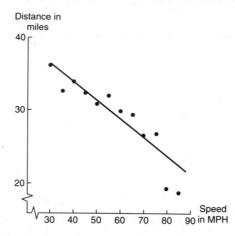

(b) Draw the line that you think fits best (specific equation in Sec. 10.2).
(c) Moderate

Section 10.2

Note: Results will vary slightly depending on rounding at intermediate steps.

2. (a & c)   Percentage of 16- to 19-Year-Olds Not in School and Per Capita Income (Thousands of Dollars)

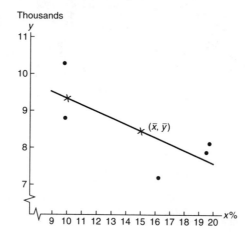

(b) $\bar{x} = 15.02$; $\bar{y} = 8.46$; $b = -0.175838$; $y = -0.176x + 11.10$
(d) $S_e = 0.919816$ (e) 8.11   (f) 6.65 to 9.57

4. (a & c) List Price and Best Price for Models of Ford Ranger (Thousands of Dollars)

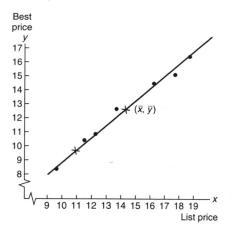

(b) $\bar{x} = 14.286$; $\bar{y} = 12.614$; $b = 0.862903$; $y = 0.863x + 0.29$
(d) $S_e = 0.348063$ (e) 13.2   (f) 12.5 to 14.0

6. (a & c) Percent Change in Wages and Percent Change in Consumer Prices

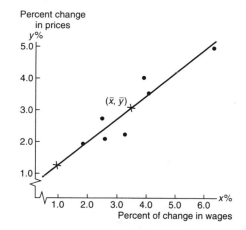

(b) $\bar{x} = 3.514$; $\bar{y} = 3.043$; $b = 0.708436$; $y = 0.708x + 0.55$
(d) $S_e = 0.487613$ (e) 4.1   (f) 3.3 to 4.9

8. (a & c) Number of Research Programs and Mean Number of Patents per Program

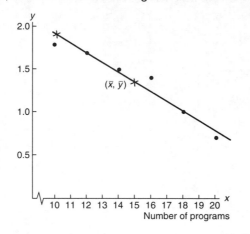

(b) $\bar{x} = 15.0$; $\bar{y} = 1.35$; $b = -0.1100$; $y = -0.11x + 3.0$
(d) $S_e = 0.109545$ (e) 1.35   (f) 1.14 to 1.56

10. (a & c) Ages of Children and Their Responses to Questions

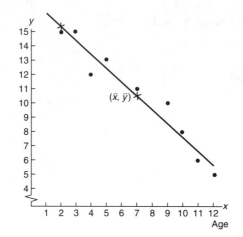

(b) $\bar{x} = 7.0$; $\bar{y} = 10.56$; $b = -0.962963$; $y = -0.963x + 17.30$
(d) $S_e = 0.931518$ (e) 8.15   (f) 4.6 to 11.7

12. (a & c) Fouls and Basketball Losses

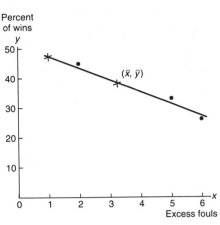

(b) $\bar{x} = 3.25$; $\bar{y} = 38.5$; $b = -3.934066$; $y = -3.934x + 51.29$
(d) $S_e = 2.109632$ (e) 35.55 (f) 31.06 to 40.04

14. (a & c) Years of Experience and Annual Salary for Public Relations Director

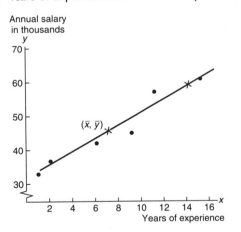

(b) $\bar{x} = 7.33$; $\bar{y} = 45.83$; $b = 1.990826$; $y = 1.991x + 31.23$
(d) $S_e = 3.034016$ (e) 47.16 (f) 40.16 to 54.16

16. (a) Result checks. (b) Result checks. (c) Yes
(d) The equation $x = 0.9337y - 0.1335$ does not match part b.
(e) In general, switching x and y values produced a *different* least-squares equation. It is important that when you perform a linear regression you know which variable is the explanatory variable and which is the response variable.

2. (a) No
   (b) Increase in buying power due to increase in salaries
4. (a) No
   (b) Increase in population could account for increases in both consumption of soda pop and in number of traffic accidents.

6. (a) Per Capita Income and Death Rates in Small Cities in Oregon

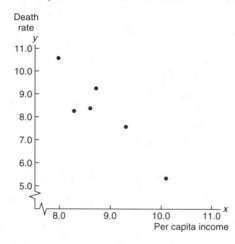

   (b) −1
   (c) $r = 0.919$; $r^2 = 0.845$; 84.5% explained; 15.5% unexplained

8. (a) Percentage of 16- to 19-Year-Olds Not in School and Death Rate per 1000 Residents

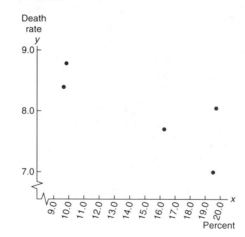

   (b) −1
   (c) $r = -0.778$; $r^2 = 0.605$; 60.5% explained; 39.5% unexplained

10. (a) Drivers' Age and Fatal Accident Rate Due to Not Yielding

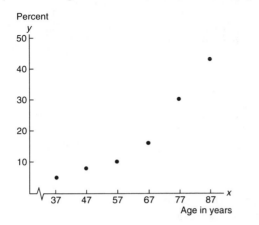

(b) 1
(c) $r = 0.943$; $r^2 = 0.889$; 88.9% explained; 11.1% unexplained

12. (a) List Price and Best Price for Models of Dodge Dakota (Thousands of Dollars)

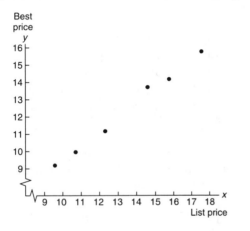

(b) 1
(c) $r = 0.9964$; $r^2 = 0.993$; 99.3% explained; 0.7% unexplained

14. (a) Time to Respond and Number of Words Used

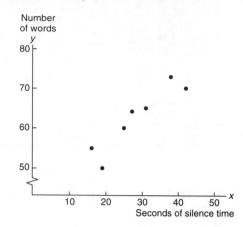

(b) 1
(c) $r = 0.9247$; $r^2 = 0.885$; 85.5% explained; 14.5% unexplained
16. (a) $SS_{xy} = SS_{yx}$ (b) Same (c) Same
   (d) $r = 0.618590$ in both cases; least-squares equations are not necessarily the same.

Section 10.4

2. $H_0: \rho = 0$; $H_1: \rho > 0$; $r_0 = 0.44$; $r = 0.646$; reject H_0
4. $H_0: \rho = 0$; $H_1: \rho > 0$; $r_0 = 0.54$; $r = 0.737$; reject H_0
6. $H_0: \rho = 0$; $H_1: \rho > 0$; $r_0 = 0.93$; $r = 0.999$; reject H_0
8. $H_0: \rho = 0$; $H_1: \rho \neq 0$; $\pm r_0 = \pm 0.87$; $r = 0.412$; fail to reject H_0
10. $H_0: \rho = 0$; $H_1: \rho < 0$; $r_0 = -0.75$; $r = -0.60$; fail to reject H_0
12. $H_0: \rho = 0$; $H_1: \rho > 0$; $r_0 = 0.48$; $r = 0.71$; reject H_0

Section 10.5

2. (a) Response variable is x_3; explanatory variables are x_1, x_4, x_7.
   (b) Constant term is -16.5; 4.0 with x_1; 9.2 with x_4; -1.1 with x_7.
   (c) 12.1   (d) 9.2 units; 27.6 units; -18.4 units
   (e) 7.55 to 10.85
   (f) $H_0: \beta_4 = 0$; $H_1: \beta_4 \neq 0$; $t_0 = \pm 3.106$; $t = 9.989$; reject H_0
4. See computer printout.
6. See computer printout.

2. (a & c) Annual Salary (Thousands) and Number of Job Changes

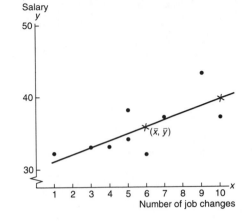

(b) $\bar{x} = 6.0$; $\bar{y} = 35.9$; $b = 0.939024$; $y = 0.939x + 30.266$
(d) 32.14   (e) $S_e = 2.564058$ (f) 26.72 to 37.57   (g) Positive
(h) $r = 0.761$; $r^2 = 0.579$
(i) H_0: $\rho = 0$; H_1: $\rho > 0$; $r_0 = 0.54$; reject H_0

4. (a & c) Number of Insurance Sales and Number of Visits

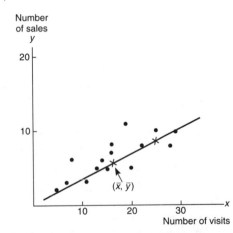

(b) $\bar{x} = 16.53$; $\bar{y} = 6.47$; $b = 0.292784$; $y = 0.293x + 1.626$
(d) $S_e = 1.730940$ (e) 6.896   (f) 3.73 to 10.07
(g) $r = 0.79$; $r^2 = 0.624$
(h) H_0: $\rho = 0$; H_1: $\rho > 0$; $r_0 = 0.59$; reject H_0

6. (a & c) List Price and Best Price for 1994 Ford Ranger

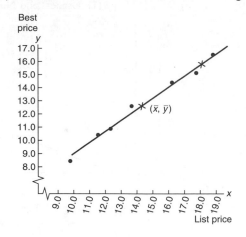

(b) High
(c) $\bar{x} = 14.286$; $\bar{y} = 12.614$; $b = 0.862903$; $y = 0.863x + 0.29$
(d) $r = 0.994$; $r^2 = 0.988$; 98.8% explained; 1.2% unexplained
(e) 13.226
(f) $S_e = 0.348063$; 12.47 to 13.98
(g) H_0: $\rho = 0$; H_1: $\rho > 0$; $r_0 = 0.67$; reject H_0

Chapter 11

Section 11.1

2. H_0: Anxiety level and need to succeed are independent.
   H_1: Anxiety level and need to succeed are not independent.
   $\chi^2 = 29.60$; $\chi^2_{0.01} = 13.28$; reject H_0, and conclude they are not independent.
4. H_0: Type of driving preferred and sex are independent.
   H_1: Type of driving preferred and sex are not independent.
   $\chi^2 = 0.017$; $\chi^2_{0.01} = 9.21$; fail to reject H_0; they are independent.
6. H_0: Amount given and ethnic group are independent.
   H_1: Amount given and ethnic group are not independent.
   $\chi^2 = 190.47$; $\chi^2_{0.01} = 26.22$; reject H_0; they are not independent.
8. H_0: Hours given and type of volunteer are independent.
   H_1: Hours given and type of volunteer are not independent.
   $\chi^2 = 16.69$; $\chi^2_{0.05} = 9.49$; reject H_0; they are not independent.
10. H_0: Stone tool construction material is independent of the site.
    H_1: Stone tool construction material is not independent of the site.
    $\chi^2 = 609.84$; $\chi^2_{0.01} = 11.34$; reject H_0; they are not independent.

Section 11.2

2. H_0: The distributions are the same. H_1: The distributions are different. $\chi^2 = 0.95$; $\chi^2_{0.01} = 11.34$; fail to reject H_0; the distributions are the same, so the claim made by Jimmy Nuts is true.

4. H_0: The distributions are the same. H_1: The distributions are different. χ^2 = 9.33; $\chi^2_{0.05}$ = 7.81; reject H_0; the distribution of fish did change.
6. H_0: The distribution of books in the library fits the distribution of books checked out. H_1: The distribution of books in the library is different from the distribution of books checked out. χ^2 = 91.51; $\chi^2_{0.05}$ = 9.49; reject H_0; the distributions are not the same.
8. H_0: The distributions are the same. H_1: The distributions are different. χ^2 = 0.35; $\chi^2_{0.05}$ = 9.49; fail to reject H_0; the distributions are the same.
10. H_0: The distributions are the same. H_1: The distributions are different. χ^2 = 15.65; $\chi^2_{0.01}$ = 15.09; reject H_0; the distributions are different.

Section 11.3

2. (a) H_0: $\sigma^2 = 23.04$; H_1: $\sigma^2 > 23.04$; critical value is 76.15. Since the observed value 80.75 falls in the critical region, we reject H_0 and conclude that the new rivets have higher variance.
   (b) $\chi^2_U = 67.50$; $\chi^2_L = 34.76$; $27.56 < \sigma^2 < 53.52$
4. H_0: $\sigma^2 = 0.007$; H_1: $\sigma^2 > 0.007$; critical value is 40.11. Since the observed value 61.71 falls in the critical region, we reject H_0. The variance is larger than 0.007, and the machine is out of control.
6. (a) H_0: $\sigma^2 = 225$; H_1: $\sigma^2 > 225$; critical value is 21.67. Since the observed value 23.04 is in the critical region, we reject H_0. The variance is larger than specified.
   (b) $\chi^2_U = 19.02$; $\chi^2_L = 2.70$; $272.56 < \sigma^2 < 1920.0$
   (c) $16.51 < \sigma < 43.82$
8. (a) H_0: $\sigma^2 = 5625$; H_1: $\sigma^2 \neq 5625$; critical values are 44.18 and 9.26. Since the observed value 21.20 falls outside the critical region, we fail to reject H_0.
   (b) $\chi^2_U = 44.18$; $\chi^2_L = 9.26$; $2698.78 < \sigma^2 < 12876.03$
   (c) $51.95 < \sigma < 113.47$

Section 11.4

2. H_0: $\mu_1 = \mu_2 = \mu_3 = \mu_4$; H_1: Not all the means are equal.

Source of Variation	Sum of Squares	Degrees of Freedom	MS	F Ratio	Critical Value	Test Decision
Between groups	18.965	3	6.322	14.910	3.41	Reject H_0
Within groups	5.517	13	0.424			
Total	24.482	16				

4. H_0: $\mu_1 = \mu_2 = \mu_3 = \mu_4$; H_1: Not all the means are equal.

Source of Variation	Sum of Squares	Degrees of Freedom	MS	F Ratio	Critical Value	Test Decision
Between groups	2.848	3	0.949	2.172	5.29	Fail to reject H_0
Within groups	6.998	16	0.437			
Total	9.846	19				

6. H_0: $\mu_1 = \mu_2 = \mu_3$; H_1: Not all the means are equal.

Source of Variation	Sum of Squares	Degrees of Freedom	MS	F Ratio	Critical Value	Test Decision
Between groups	26.00	2	13.00	0.471	4.26	Fail to reject H_0
Within groups	248.25	9	27.58			
Total	274.25	11				

8. H_0: $\mu_1 = \mu_2 = \mu_3 = \mu_4 = \mu_5$; H_1: Not all the means are equal.

Source of Variation	Sum of Squares	Degrees of Freedom	MS	F Ratio	Critical Value	Test Decision
Between groups	0.371	4	0.093	1.209	5.99	Fail to reject H_0
Within groups	0.767	10	0.077			
Total	1.138	14				

Chapter Review Problems

2. H_0: Time to do a test and test score are independent. H_1: Time to do a test and test score are not independent. $\chi^2 = 3.92$; $\chi^2_{0.01} = 11.34$. Do not reject H_0; time to do a test and test results are independent.

4. H_0: $\mu_1 = \mu_2 = \mu_3$; H_1: Not all the means are equal.

Source of Variation	Sum of Squares	Degrees of Freedom	MS	F Ratio	Critical Value	Test Decision
Between groups	1.002	2	0.501	0.443	8.02	Fail to reject H_0
Within groups	10.165	9	1.129			
Total	11.167	11				

6. H_0: $\sigma^2 = 0.0625$; H_1: $\sigma^2 > 0.0625$; critical value $\chi^2_{0.05} = 19.68$. Since the observed value $\chi^2 = 25.41$ falls in the critical region, we reject H_0; the machine needs to be adjusted.

Chapter 12

Section 12.1

2. H_0: $\mu_1 = \mu_2$; H_1: $\mu_1 > \mu_2$; $z_0 = 1.645$; $z = 0.83$ for $r = 8/13$; fail to reject H_0
4. H_0: $\mu_1 = \mu_2$; H_1: $\mu_1 \neq \mu_2$; $\pm z_0 = \pm 2.58$; $z = 0.53$ for $r = 8/14$; fail to reject H_0
6. H_0: $\mu_1 = \mu_2$; H_1: $\mu_1 < \mu_2$; $z_0 = -1.645$; $z = -0.69$ for $r = 8/19$; fail to reject H_0
8. H_0: $\mu_1 = \mu_2$; H_1: $\mu_1 < \mu_2$; $z_0 = -1.645$; $z = -2.32$ for $r = 3/15$; reject H_0
10. H_0: $\mu_1 = \mu_2$; H_1: $\mu_1 \neq \mu_2$; $\pm z_0 = \pm 1.96$; $z = 1.39$ for $r = 9/13$; fail to reject H_0

Section 12.2

2. H_0: $\mu_1 = \mu_2$; H_1: $\mu_1 \neq \mu_2$; $\pm z_0 = \pm 1.96$; $\mu_R = 115$; $\sigma_R = 15.17$; $R = 117$; $z = 0.13$; fail to reject H_0
4. H_0: $\mu_1 = \mu_2$; H_1: $\mu_1 \neq \mu_2$; $\pm z_0 = \pm 1.96$; $\mu_R = 105$; $\sigma_R = 13.23$; $R = 92$; $z = -0.98$; fail to reject H_0
6. H_0: $\mu_1 = \mu_2$; H_1: $\mu_1 \neq \mu_2$; $\pm z_0 = \pm 2.58$; $\mu_R = 110$; $\sigma_R = 14.20$; $R = 97.5$; $z = -0.88$; fail to reject H_0
8. H_0: $\mu_1 = \mu_2$; H_1: $\mu_1 \neq \mu_2$; $\pm z_0 = \pm 2.58$; $\mu_R = 85.5$; $\sigma_R = 11.32$; $R = 99$; $z = 1.19$; fail to reject H_0
10. H_0: $\mu_1 = \mu_2$; H_1: $\mu_1 \neq \mu_2$; $\pm z_0 = \pm 2.58$; $\mu_R = 68$; $\sigma_R = 9.52$; $R = 75.5$; $z = 0.79$; fail to reject H_0

Section 12.3

2. H_0: $\rho_S = 0$; H_1: $\rho_S \neq 0$; critical values 0.680 and -0.680. Since the observed value $r_S = 0.349$ falls outside the critical region, we do not reject H_0; there is no monotone relation between costs and earnings.
4. H_0: $\rho_S = 0$; H_1: $\rho_S > 0$; critical value is 0.564. Since $r_S = 0.488$ is outside the critical region, we do not reject H_0; there is no monotone relation between rank of finish and rank of score.

6. H_0: $\rho_S = 0$; H_1: $\rho_S \neq 0$; critical values are 0.715 and -0.715. Since the observed value $r_S = 0.452$ falls outside the critical region, we do not reject H_0; there is no monotone relation between quality rank and price rank.

8. H_0: $\rho_S = 0$; H_1: $\rho_S > 0$; critical value is 0.829. Since the observed value $r_S = 0.257$ falls outside the critical region, we fail to reject H_0; there is no monotone relation between the opinions of the managers.

Chapter Review Problems

2. H_0: $\mu_1 = \mu_2$; H_1: $\mu_1 > \mu_2$; $z_0 = 1.645$; $z = 2.17$ for $r = 0.79$; reject H_0

4. H_0: $\mu_1 = \mu_2$; H_1: $\mu_1 \neq \mu_2$; $\pm z_0 = \pm 1.96$; $\mu_R = 90$; $\sigma_R = 12.25$; $R = 116$; $z = 2.12$; reject H_0

6. H_0: $\rho_S = 0$; H_1: $\rho_S \neq 0$; critical values are 0.900 and -0.900. Since the observed value $r_S = -0.800$ falls in the critical region, we reject H_0; there is a monotone relation between opinions of the chefs.

Part IV

Transparency Masters

Figure 2-17
Minitab Generated

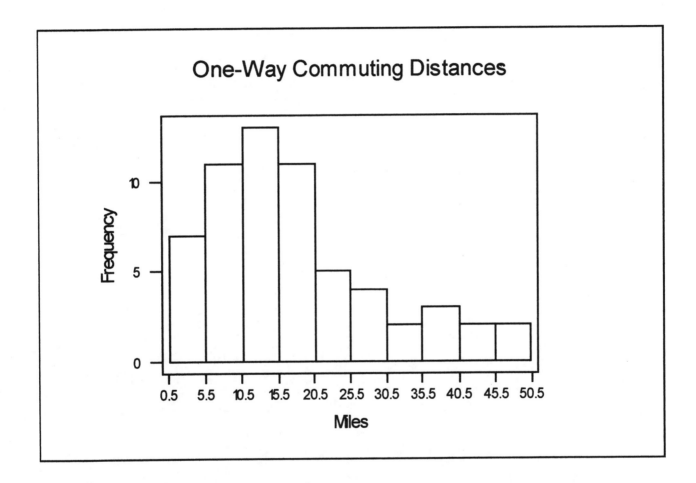

Figure 2-30
Minitab Generated

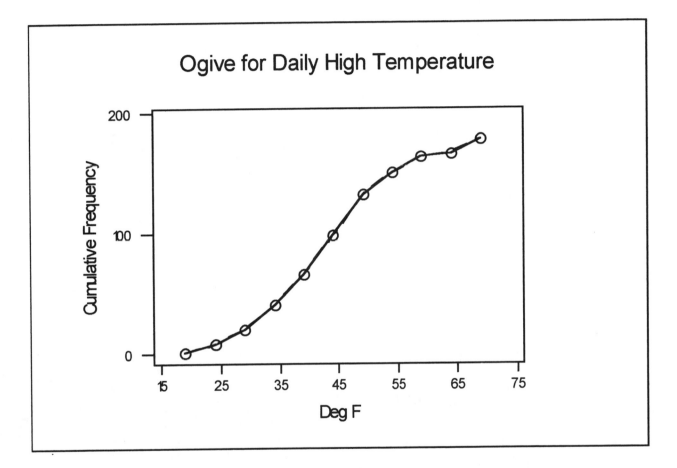

Ogive for Daily High Temperature

Figure 2-37
Weight of Carry-On Luggage

Leaf Unit = 1 Pound

3| 2 represents 32

Stem	Leaf
0	3 0
1	2 7 8 8 9 2 8
2	7 7 2 9 1 6 1 8 9 1 6
3	0 5 8 6 5 8 2 3 2 1 2 3 1 2
4	2 7 1 5 3
5	1

Box-and-Whisker Plot

Figure 3-14
TI-82 Generated

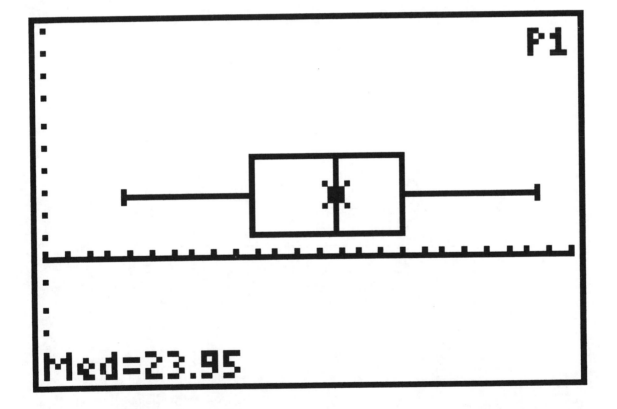

Sample Space for Throwing Two Dice

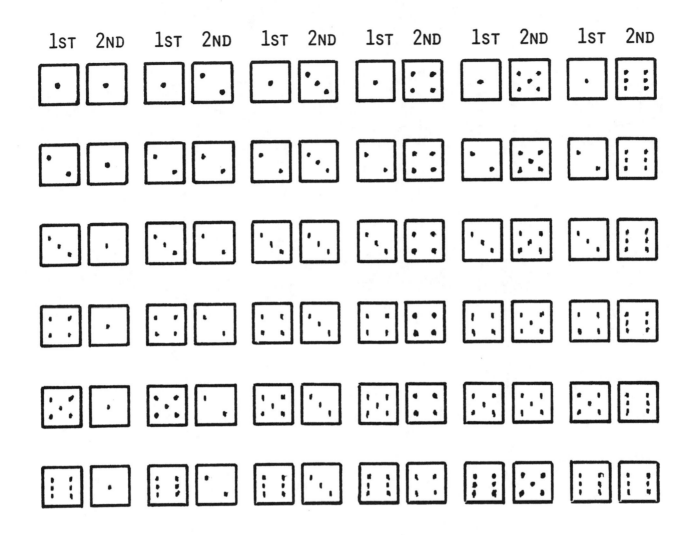

Sample Space for Drawing One Card
From a Standard Bridge Deck

	Red Cards		Black Cards	
Hearts	Diamonds	Clubs	Spades
2	2	2	2
3	3	3	3
4	4	4	4
5	5	5	5
6	6	6	6
7	7	7	7
8	8	8	8
9	9	9	9
10	10	10	10
Jack	Jack	Jack	Jack
Queen	Queen	Queen	Queen
King	King	King	King
Ace	Ace	Ace	Ace

Figure 6-7
Areas Under a Normal Curve

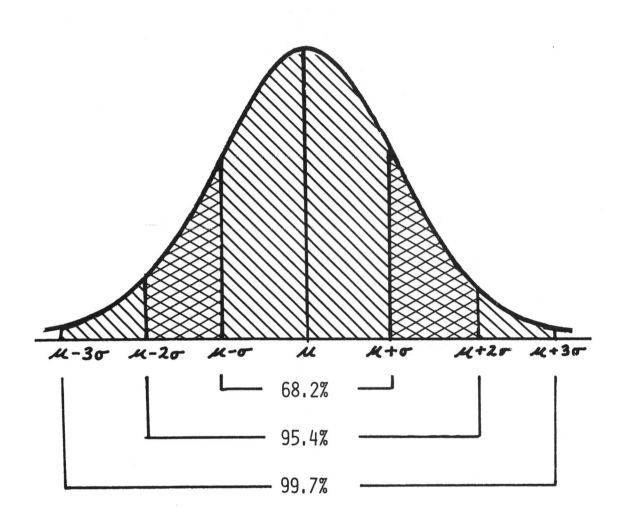

Figure 6-11
Control Chart for Number of Rooms Not Made Up
ComputerStat Generated

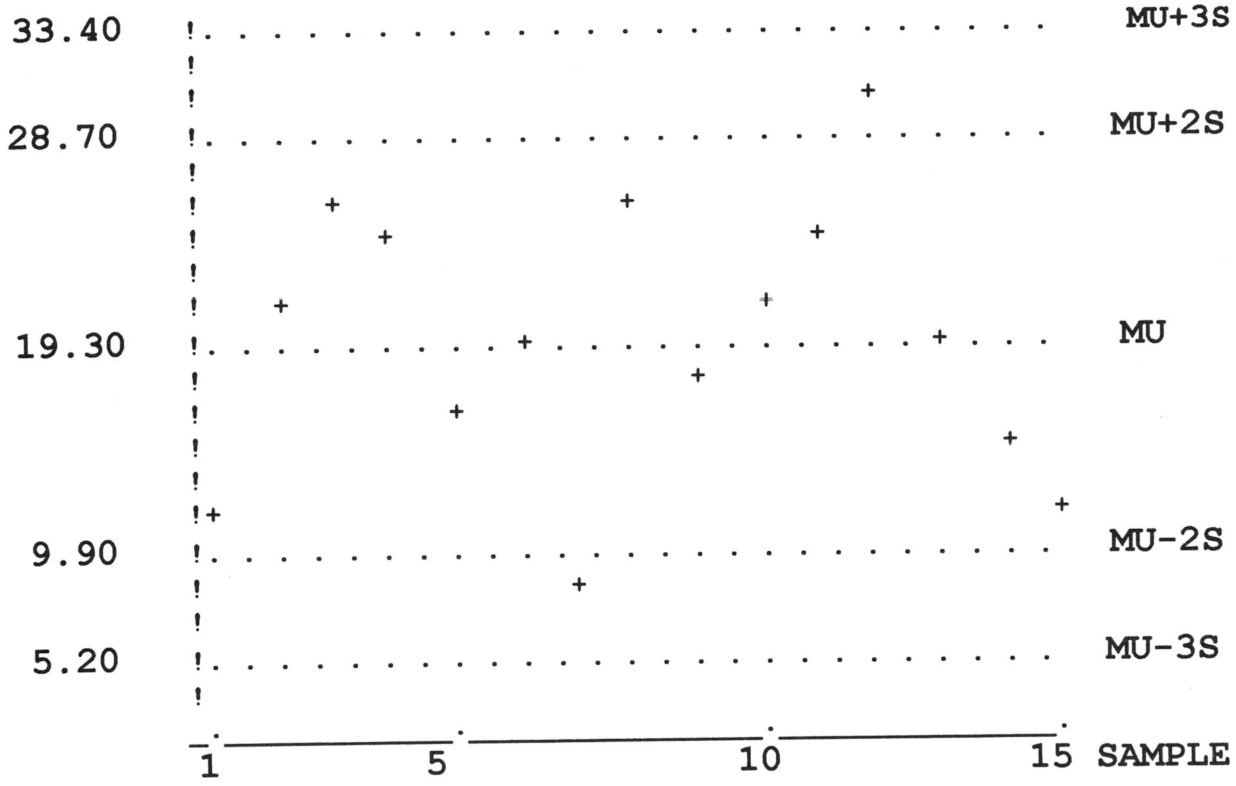

```
33.40  !. . . . . . . . . . . . . . .  MU+3S
       !
       !                         +
28.70  !. . . . . . . . . . . . . . .  MU+2S
       !
       !      +           +
       !         +              +
       !
       !   +              +
19.30  !. . . . . + . . . . . . + . . .  MU
       !                  +
       !      +
       !                           +
       !
       !+                       +
 9.90  !. . . . . . . . . . . . . . .  MU-2S
       !                +
       !
 5.20  !. . . . . . . . . . . . . . .  MU-3S
       !
       ._____._____._____.___
        1       5        10       15  SAMPLE
```

Section 9.4 Problem #7
ComputerStat Display

****** INFORMATION SUMMARY ******
WE USE A T-DISTRIBUTION WITH d.f.= 5

SAMPLE SIZE N = 6
SAMPLE MEAN M = 4.066667
SAMPLE STANDARD DEVIATION S = .4412104

LEVEL OF SIGNIFICANCE ALPHA = .05

NULL HYPOTHESIS H0:MU = 4.8
ALTERNATE HYPOTHESIS H1:MU < 4.8
TYPE OF TEST - LEFT TAIL

P-VALUE = 4.810899E-03

SHALL WE ACCEPT OR REJECT THE NULL HYPOTHESIS ?
COMPARE ALPHA AND THE P-VALUE AS LISTED ABOVE

MAKE YOUR DECISION , THEN PRESS ANY KEY TO SEE
   THE RESULTS

   ****** TEST CONCLUSION ******

1) SINCE   P-VALUE = 4.810899E-03 <= ALPHA = .05
   ***** WE REJECT THE NULL HYPOTHESIS *****

2) USING CRITICAL REGIONS, WE GET THE SAME RESULT
   THE CRITICAL REGION IS ALL T-VALUES <= -2.015552
   THE SAMPLE TEST STATISTIC T=-4.071284
   IS INSIDE THE CRITICAL REGION, THEREFORE
   ***** WE REJECT THE NULL HYPOTHESIS *****

Least Squares Line and Scatter Diagram
Figure 10-9

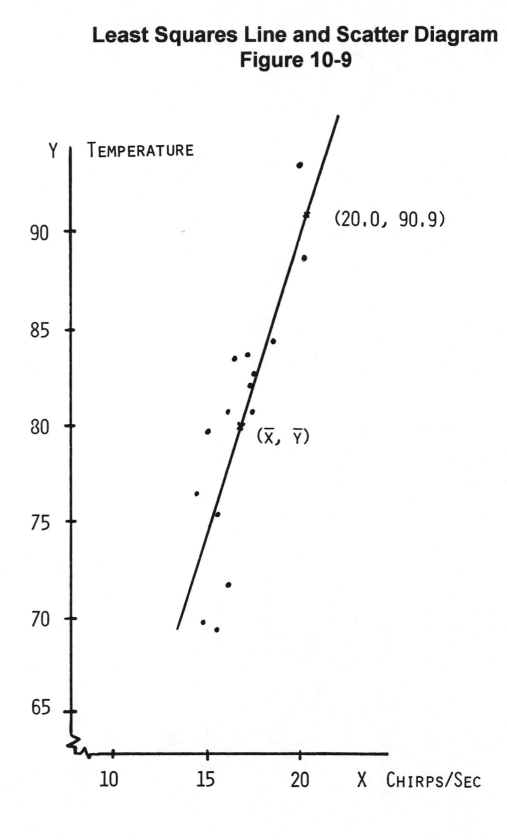

Least Squares Line and Scatter Diagram
Figure 10-9

Figure 10-15
95% Confidence Band for Predicted Values
MINITAB Generated

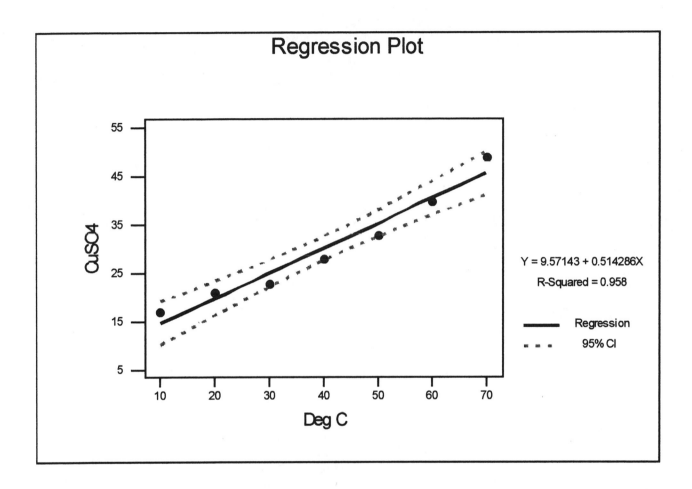

Regression Plot

$Y = 9.57143 + 0.514286X$

R-Squared = 0.958

—— Regression

∙ ∙ ∙ 95% CI

Part V

Formula Card Masters

Masters for Statistical Tables

Frequently Used Formulas

n = sample size N = population size f = frequency

Chapter 2

Class Width = $\dfrac{\text{high} - \text{low}}{2}$ (increase to next integer)

Class Midpoint = $\dfrac{\text{upper limit} - \text{lower limit}}{2}$

Lower boundary = lower boundary of previous class + class width

Chapter 3

Sample mean $\bar{x} = \dfrac{\Sigma x}{n}$

Population mean $\mu = \dfrac{\Sigma x}{N}$

Weighted average = $\dfrac{\Sigma xw}{\Sigma w}$

Range = largest data value − smallest data value

Sample standard deviation $s = \sqrt{\dfrac{\Sigma (x - \bar{x})^2}{n - 1}}$

Computation formula $s = \sqrt{\dfrac{SS_x}{n - 1}}$ where

$SS_x = \Sigma x^2 - \dfrac{(\Sigma x)^2}{n}$

Population standard deviation $\sigma = \sqrt{\dfrac{\Sigma (x - \mu)^2}{N}}$

Sample variance s^2

Population variance σ^2

Sample Coefficient of Variation $CV = \dfrac{s}{\bar{x}} \cdot 100$

Sample mean for grouped data $\bar{x} = \dfrac{\Sigma xf}{n}$

Sample standard deviation for grouped data

$s = \sqrt{\dfrac{\Sigma (x - \bar{x})^2 f}{n - 1}}$

Chapter 4

Probability of the complement of event A
$P(\text{not } A) = 1 - P(A)$

Multiplication rule for independent events
$P(A \text{ and } B) = P(A) \cdot P(B)$

General multiplication rules
$P(A \text{ and } B) = P(A) \cdot P(B, \text{given } A)$
$P(A \text{ and } B) = P(B) \cdot P(A, \text{given } B)$

Addition rule for mutually exclusive events
$P(A \text{ or } B) = P(A) + P(B)$

General addition rule
$P(A \text{ or } B) = P(A) + P(B) - P(A \text{ and } B)$

Permutation rule $P_{n,r} = \dfrac{n!}{(n - r)!}$

Combination rule $C_{n,r} = \dfrac{n!}{r!(n - r)!}$

Mean of a discrete probability distribution $\mu = \Sigma x P(x)$

Standard deviation of a discrete probability distribution
$\sigma = \sqrt{\Sigma (x - \mu)^2 P(x)}$

Chapter 5

For Binomial Distributions

r = number of successes; p = probability of success;
$q = 1 - p$

Binomial probability distribution $P(r) = \dfrac{n!}{r!(n - r)!} p^r q^{n-r}$

Mean $\mu = np$

Standard deviation $\sigma = \sqrt{npq}$

Geometric Probability Distribution

n = number of trial on which first success occurs
$P(n) = p(1 - p)^{n-1}$

Poisson Probability Distribution

λ = mean number of successes over given interval
$P(\lambda) = \dfrac{e^{-\lambda} \lambda^r}{r!}$

Chapter 6

Raw score $x = z\sigma + \mu$

Standard score $z = \dfrac{x - \mu}{\sigma}$

Chapter 7

Mean of $\bar{x}$ distribution $\mu_{\bar{x}} = \mu$

Standard deviation of $\bar{x}$ distribution $\sigma_{\bar{x}} = \dfrac{\sigma}{\sqrt{n}}$

Standard score for $\bar{x}$ $\quad z = \dfrac{\bar{x} - \mu}{\sigma/\sqrt{n}}$

Chapter 8

Confidence Interval

for $\mu(n \geq 30)$

$$\bar{x} - z_c \frac{\sigma}{\sqrt{n}} < \mu < \bar{x} + z_c \frac{\sigma}{\sqrt{n}}$$

for $\mu(n < 30)$

$$d.f. = n - 1$$

$$\bar{x} - t_c \frac{s}{\sqrt{n}} < \mu < \bar{x} + t_c \frac{s}{\sqrt{n}}$$

for $p(np > 5 \text{ and } nq > 5)$

$$\hat{p} - z_c \sqrt{\frac{\hat{p}(1 - \hat{p})}{n}} < p < \hat{p} + z_c \sqrt{\frac{\hat{p}(1 - \hat{p})}{n}} \text{ where } \hat{p} = r/n$$

for difference of means $(n_1 \geq 30 \text{ and } n_2 \geq 30)$

$$(\bar{x}_1 - \bar{x}_2) - z_c \sqrt{\frac{\sigma_1^2}{n_1} + \frac{\sigma_2^2}{n_2}} < \mu_1 - \mu_2 < (\bar{x}_1 - \bar{x}_2)$$

$$+ z_c \sqrt{\frac{\sigma_1^2}{n_1} + \frac{\sigma_2^2}{n_2}}$$

for difference of means $(n_1 < 30 \text{ and/or } n_2 < 30 \text{ and } \sigma_1 \approx \sigma_2)$

$$d.f. = n_1 + n_2 - 2$$

$$(\bar{x} - \bar{x}_2) - t_c s \sqrt{\frac{1}{n_1} + \frac{1}{n_2}} < \mu_1 - \mu_2 < (\bar{x}_1 - \bar{x}_2)$$

$$+ t_c s \sqrt{\frac{1}{n_1} + \frac{1}{n_2}}$$

$$\text{where } s = \sqrt{\frac{(n_1 - 1)s_1^2 + (n_2 - 1)s_2^2}{n_1 + n_2 - 2}}$$

for difference of proportions

where $\hat{p}_1 = r_1/n_1; \hat{p}_2 = r_2/n_2; \hat{q}_1 = 1 - \hat{p}_1; \hat{q}_2 = 1 - \hat{p}_2$

$$(\hat{p}_1 - \hat{p}_2) - z_c \sqrt{\frac{\hat{p}_1 \hat{q}_1}{n_1} + \frac{\hat{p}_2 \hat{q}_2}{n_2}} < p_1 - p_2 < (\hat{p}_1 - \hat{p}_2)$$

$$+ z_c \sqrt{\frac{\hat{p}_1 \hat{q}_1}{n_1} + \frac{\hat{p}_2 \hat{q}_2}{n_2}}$$

Sample Size for Estimating

means $n = \left(\dfrac{z_c \sigma}{E}\right)^2$

proportions

$n = p(1 - p)\left(\dfrac{z_c}{E}\right)^2$ with preliminary estimate for p

$n = \dfrac{1}{4}\left(\dfrac{z_c}{E}\right)^2$ without preliminary estimate for p

Chapter 9

Sample Test Statistics for Tests of Hypotheses

for $\mu(n \geq 30)$ $\quad z = \dfrac{\bar{x} - \mu}{\sigma/\sqrt{n}}$

for $\mu(n < 30); d.f. = n - 1$ $\quad t = \dfrac{\bar{x} - \mu}{s/\sqrt{n}}$

for p $\quad z = \dfrac{\hat{p} - p}{\sqrt{pq/n}}$

for paired differences d $\quad t = \dfrac{\bar{d} - \mu_d}{s_d/\sqrt{n}}$ with $d.f. = n - 1$

difference of means large sample

$$z = \frac{(\bar{x}_1 - \bar{x}_2) - (\mu_1 - \mu_2)}{\sqrt{\dfrac{\sigma_1^2}{n_1} + \dfrac{\sigma_2^2}{n_2}}}$$

difference of means small sample with $\sigma_1 \approx \sigma_2$;

$$d.f. = n_1 + n_2 - 2$$

$$t = \frac{(\bar{x}_1 - \bar{x}_2) - (\mu_1 - \mu_2)}{s \sqrt{\dfrac{1}{n_1} + \dfrac{1}{n_2}}}$$

where $s = \sqrt{\dfrac{(n_1 - 1)s_1^2 + (n_2 - 1)s_2^2}{n_1 + n_2 - 2}}$

difference of proportions

$$z = \frac{\hat{p}_1 - \hat{p}_2}{\sqrt{\dfrac{\bar{p}\bar{q}}{n_1} + \dfrac{\bar{p}\bar{q}}{n_2}}} \text{ where } \bar{p} = \frac{r_1 + r_2}{n_1 + n_2}; \bar{q} = 1 - \bar{p};$$

$$\hat{p}_1 = r_1/n_1; \hat{p}_2 = r_2/n_2$$

Chapter 10

Regression and Correlation

In all these formulas

$$SS_x = \Sigma x^2 - \frac{(\Sigma x)^2}{n}$$

$$SS_y = \Sigma y^2 - \frac{(\Sigma y)^2}{n}$$

$$SS_{xy} = \Sigma xy - \frac{(\Sigma x)(\Sigma y)}{n}$$

Least squares line $y = a + bx$ where $b = \dfrac{SS_{xy}}{SS_x}$ and
$a = \bar{y} - b\bar{x}$,

Standard error of estimate $S_e = \sqrt{\dfrac{SS_y - bSS_{xy}}{n - 2}}$
 where $b = \dfrac{SS_{xy}}{SS_x}$

Pearson product-moment correlation coefficient

$$r = \frac{SS_{xy}}{\sqrt{SS_x SS_y}}$$

Coefficient of determination $= r^2$

Confidence interval for y
 $y_p - E < y < y_p + E$ where y_p is the predicted y value
 for x

$$E = t_e S_e \sqrt{1 + \frac{1}{n} + \frac{(x - \bar{x})^2}{SS_x}}$$

Chapter 11

$$\chi^2 = \Sigma \frac{(O - E)^2}{E} \quad \text{where } E = \frac{\text{(row total)(column total)}}{\text{sample size}}$$

Tests of Independence $d.f. = (R - 1)(C - 1)$

Goodness of fit $d.f. = \text{(number of entries)} - 1$

Confidence Interval for σ^2; $d.f. = n - 1$

$$\frac{(n - 1)s^2}{\chi_U^2} < \sigma^2 < \frac{(n - 1)s^2}{\chi_L^2}$$

Sample test statistic for H_0: $\sigma^2 = k$; $d.f. = n - 1$

$$\chi^2 = \frac{(n - 1)s^2}{\sigma^2}$$

ANOVA

k = number of groups; N = total sample size

$$SS_{TOT} = \Sigma x_{TOT}^2 - \frac{(\Sigma x_{TOT})^2}{N}$$

$$SS_{BET} = \sum_{all\ groups} \left(\frac{(\Sigma x_i)^2}{n_i} \right) - \frac{(\Sigma x_{TOT})^2}{N}$$

$$SS_W = \sum_{all\ groups} \left(\Sigma x_i^2 - \frac{(\Sigma x_i)^2}{n_i} \right)$$

$$SS_{TOT} = SS_{BET} + SS_W$$

$$MS_{BET} = \frac{SS_{BET}}{d.f._{BET}} \quad \text{where } d.f._{BET} = k - 1$$

$$MS_W = \frac{SS_W}{d.f._W} \quad \text{where } d.f._W = N - k$$

$$F = \frac{MS_{BET}}{MS_W} \quad \text{where } d.f. \text{ numerator} = d.f._{BET} = k - 1;$$
 $d.f.$ denominator $= d.f._W = N - k$

Chapter 12

Sample test statistic for r = proportion of plus signs to all
 signs ($n \geqslant 12$)

$$z = \frac{r - 0.5}{\sqrt{0.25/n}}$$

Sample test statistic for R = sum of ranks

$$z = \frac{R - \mu_R}{\sigma_R} \quad \text{where } \mu_R = \frac{n_1(n_1 + n_2 + 1)}{2} \text{ and}$$

$$\sigma_R = \sqrt{\frac{n_1 n_2(n_1 + n_2 + 1)}{12}}$$

Spearman rank correlation coefficient

$$r_s = 1 - \frac{6\Sigma d^2}{n(n^2 - 1)} \quad \text{where } d = x - y$$

Table I Random Numbers

92630	78240	19267	95457	53497	23894	37708	79862	76471	66418
79445	78735	71549	44843	26104	67318	00701	34986	66751	99723
59654	71966	27386	50004	05358	94031	29281	18544	52429	06080
31524	49587	76612	39789	13537	48086	59483	60680	84675	53014
06348	76938	90379	51392	55887	71015	09209	79157	24440	30244
28703	51709	94456	48396	73780	06436	86641	69239	57662	80181
68108	89266	94730	95761	75023	48464	65544	96583	18911	16391
99938	90704	93621	66330	33393	95261	95349	51769	91616	33238
91543	73196	34449	63513	83834	99411	58826	40456	69268	48562
42103	02781	73920	56297	72678	12249	25270	36678	21313	75767
17138	27584	25296	28387	51350	61664	37893	05363	44143	42677
28297	14280	54524	21618	95320	38174	60579	08089	94999	78460
09331	56712	51333	06289	75345	08811	82711	57392	25252	30333
31295	04204	93712	51287	05754	79396	87399	51773	33075	97061
36146	15560	27592	42089	99281	59640	15221	96079	09961	05371
29553	18432	13630	05529	02791	81017	49027	79031	50912	09399
23501	22642	63081	08191	89420	67800	55137	54707	32945	64522
57888	85846	67967	07835	11314	01545	48535	17142	08552	67457
55336	71264	88472	04334	63919	36394	11196	92470	70543	29776
10087	10072	55980	64688	68239	20461	89381	93809	00796	95945
34101	81277	66090	88872	37818	72142	67140	50785	21380	16703
53362	44940	60430	22834	14130	96593	23298	56203	92671	15925
82975	66158	84731	19436	55790	69229	28661	13675	99318	76873
54827	84673	22898	08094	14326	87038	42892	21127	30712	48489
25464	59098	27436	89421	80754	89924	19097	67737	80368	08795
67609	60214	41475	84950	40133	02546	09570	45682	50165	15609
44921	70924	61295	51137	47596	86735	35561	76649	18217	63446
33170	30972	98130	95828	49786	13301	36081	80761	33985	68621
84687	85445	06208	17654	51333	02878	35010	67578	61574	20749
71886	56450	36567	09395	96951	35507	17555	35212	69106	01679
00475	02224	74722	14721	40215	21351	08596	45625	83981	63748
25993	38881	68361	59560	41274	69742	40703	37993	03435	18873
92882	53178	99195	93803	56985	53089	15305	50522	55900	43026
25138	26810	07093	15677	60688	04410	24505	37890	67186	62829
84631	71882	12991	83028	82484	90339	91950	74579	03539	90122
34003	92326	12793	61453	48121	74271	28363	66561	75220	35908
53775	45749	05734	86169	42762	70175	97310	73894	88606	19994
59316	97885	72807	54966	60859	11932	35265	71601	55577	67715
20479	66557	50705	26999	09854	52591	14063	30214	19890	19292
86180	84931	25455	26044	02227	52015	21820	50599	51671	65411
21451	68001	72710	40261	61281	13172	63819	48970	51732	54113
98062	68375	80089	24135	72355	95428	11808	29740	81644	86610
01788	64429	14430	94575	75153	94576	61393	96192	03227	32258
62465	04841	43272	68702	01274	05437	22953	18946	99053	41690
94324	31089	84159	92933	99989	89500	91586	02802	69471	68274
05797	43984	21575	09908	70221	19791	51578	36432	33494	79888
10395	14289	52185	09721	25789	38562	54794	04897	59012	89251
35177	56986	25549	59730	64718	52630	31100	62384	49483	11409
25633	89619	75882	98256	02126	72099	57183	55887	09320	73463
16464	48280	94254	45777	45150	68865	11382	11782	22695	41988

Table 2 Factorials

n	$n!$
0	1
1	1
2	2
3	6
4	24
5	120
6	720
7	5040
8	40320
9	362880
10	3628800
11	39916800
12	479001600
13	6227020800
14	87178291200
15	1307674368000
16	20922789888000
17	355687428096000
18	6402373705728000
19	121645100408832000
20	2432902008176640000

Table 3 Binomial Coefficients $C_{n,r}$

n \ r	0	1	2	3	4	5	6	7	8	9	10
1	1	1									
2	1	2	1								
3	1	3	3	1							
4	1	4	6	4	1						
5	1	5	10	10	5	1					
6	1	6	15	20	15	6	1				
7	1	7	21	35	35	21	7	1			
8	1	8	28	56	70	56	28	8	1		
9	1	9	36	84	126	126	84	36	9	1	
10	1	10	45	120	210	252	210	120	45	10	1
11	1	11	55	165	330	462	462	330	165	55	11
12	1	12	66	220	495	792	924	792	495	220	66
13	1	13	78	286	715	1,287	1,716	1,716	1,287	715	286
14	1	14	91	364	1,001	2,002	3,003	3,432	3,003	2,002	1,001
15	1	15	105	455	1,365	3,003	5,005	6,435	6,435	5,005	3,003
16	1	16	120	560	1,820	4,368	8,008	11,440	12,870	11,440	8,008
17	1	17	136	680	2,380	6,188	12,376	19,448	24,310	24,310	19,448
18	1	18	153	816	3,060	8,568	18,564	31,824	43,758	48,620	43,758
19	1	19	171	969	3,876	11,628	27,132	50,388	75,582	92,378	92,378
20	1	20	190	1,140	4,845	15,504	38,760	77,520	125,970	167,960	184,756

Table 4 Binomial Probability Distribution $C_{n,\,r}\,p^r\,q^{n-r}$

This table shows the probability of r successes in n independent trials, each with probability of success p.

n	r	.01	.05	.10	.15	.20	.25	.30	.35	.40	.45	.50	.55	.60	.65	.70	.75	.80	.85	.90	.95
2	0	.980	.902	.810	.723	.640	.563	.490	.423	.360	.303	.250	.203	.160	.123	.090	.063	.040	.023	.010	.002
	1	.020	.095	.180	.255	.320	.375	.420	.455	.480	.495	.500	.495	.480	.455	.420	.375	.320	.255	.180	.095
	2	.000	.002	.010	.023	.040	.063	.090	.123	.160	.203	.250	.303	.360	.423	.490	.563	.640	.723	.810	.902
3	0	.970	.857	.729	.614	.512	.422	.343	.275	.216	.166	.125	.091	.064	.043	.027	.016	.008	.003	.001	.000
	1	.029	.135	.243	.325	.384	.422	.441	.444	.432	.408	.375	.334	.288	.239	.189	.141	.096	.057	.027	.007
	2	.000	.007	.027	.057	.096	.141	.189	.239	.288	.334	.375	.408	.432	.444	.441	.422	.384	.325	.243	.135
	3	.000	.000	.001	.003	.008	.016	.027	.043	.064	.091	.125	.166	.216	.275	.343	.422	.512	.614	.729	.857
4	0	.961	.815	.656	.522	.410	.316	.240	.179	.130	.092	.062	.041	.026	.015	.008	.004	.002	.001	.000	.000
	1	.039	.171	.292	.368	.410	.422	.412	.384	.346	.300	.250	.200	.154	.112	.076	.047	.026	.011	.004	.000
	2	.001	.014	.049	.098	.154	.211	.265	.311	.346	.368	.375	.368	.346	.311	.265	.211	.154	.098	.049	.014
	3	.000	.000	.004	.011	.026	.047	.076	.112	.154	.200	.250	.300	.346	.384	.412	.422	.410	.368	.292	.171
	4	.000	.000	.000	.001	.002	.004	.008	.015	.026	.041	.062	.092	.130	.179	.240	.316	.410	.522	.656	.815
5	0	.951	.774	.590	.444	.328	.237	.168	.116	.078	.050	.031	.019	.010	.005	.002	.001	.000	.000	.000	.000
	1	.048	.204	.328	.392	.410	.396	.360	.312	.259	.206	.156	.113	.077	.049	.028	.015	.006	.002	.000	.000
	2	.001	.021	.073	.138	.205	.264	.309	.336	.346	.337	.312	.276	.230	.181	.132	.088	.051	.024	.008	.001
	3	.000	.001	.008	.024	.051	.088	.132	.181	.230	.276	.312	.337	.346	.336	.309	.264	.205	.138	.073	.021
	4	.000	.000	.000	.002	.006	.015	.028	.049	.077	.113	.156	.206	.259	.312	.360	.396	.410	.392	.328	.204
	5	.000	.000	.000	.000	.000	.001	.002	.005	.010	.019	.031	.050	.078	.116	.168	.237	.328	.444	.590	.774
6	0	.941	.735	.531	.377	.262	.178	.118	.075	.047	.028	.016	.008	.004	.002	.001	.000	.000	.000	.000	.000
	1	.057	.232	.354	.399	.393	.356	.303	.244	.187	.136	.094	.061	.037	.020	.010	.004	.002	.000	.000	.000
	2	.001	.031	.098	.176	.246	.297	.324	.328	.311	.278	.234	.186	.138	.095	.060	.033	.015	.006	.001	.000
	3	.000	.002	.015	.042	.082	.132	.185	.236	.276	.303	.312	.303	.276	.236	.185	.132	.082	.042	.015	.002
	4	.000	.000	.001	.006	.015	.033	.060	.095	.138	.186	.234	.278	.311	.328	.324	.297	.246	.176	.098	.031
	5	.000	.000	.000	.000	.002	.004	.010	.020	.037	.061	.094	.136	.187	.244	.303	.356	.393	.399	.354	.232
	6	.000	.000	.000	.000	.000	.001	.001	.002	.004	.008	.016	.028	.047	.075	.118	.178	.262	.377	.531	.735
7	0	.932	.698	.478	.321	.210	.133	.082	.049	.028	.015	.008	.004	.002	.001	.000	.000	.000	.000	.000	.000
	1	.066	.257	.372	.396	.367	.311	.247	.185	.131	.087	.055	.032	.017	.008	.004	.001	.000	.000	.000	.000
	2	.002	.041	.124	.210	.275	.311	.318	.299	.261	.214	.164	.117	.077	.047	.025	.012	.004	.001	.000	.000
	3	.000	.004	.023	.062	.115	.173	.227	.268	.290	.292	.273	.239	.194	.144	.097	.058	.029	.011	.003	.000
	4	.000	.000	.003	.011	.029	.058	.097	.144	.194	.239	.273	.292	.290	.268	.227	.173	.115	.062	.023	.004
	5	.000	.000	.000	.001	.004	.012	.025	.047	.077	.117	.164	.214	.261	.299	.318	.311	.275	.210	.124	.041
	6	.000	.000	.000	.000	.000	.001	.004	.008	.017	.032	.055	.087	.131	.185	.247	.311	.367	.396	.372	.257
	7	.000	.000	.000	.000	.000	.000	.000	.001	.002	.004	.008	.015	.028	.049	.082	.133	.210	.321	.478	.698

Table 4 continued

											p										
n	r	.01	.05	.10	.15	.20	.25	.30	.35	.40	.45	.50	.55	.60	.65	.70	.75	.80	.85	.90	.95
8	0	.923	.663	.430	.272	.168	.100	.058	.032	.017	.008	.004	.002	.001	.000	.000	.000	.000	.000	.000	.000
	1	.075	.279	.383	.385	.336	.267	.198	.137	.090	.055	.031	.016	.008	.003	.001	.000	.000	.000	.000	.000
	2	.003	.051	.149	.238	.294	.311	.296	.259	.209	.157	.109	.070	.041	.022	.010	.004	.001	.000	.000	.000
	3	.000	.005	.033	.084	.147	.208	.254	.279	.279	.257	.219	.172	.124	.081	.047	.023	.009	.003	.000	.000
	4	.000	.000	.005	.018	.046	.087	.136	.188	.232	.263	.273	.263	.232	.188	.136	.087	.046	.018	.005	.000
	5	.000	.000	.000	.003	.009	.023	.047	.081	.124	.172	.219	.257	.279	.279	.254	.208	.147	.084	.033	.005
	6	.000	.000	.000	.000	.001	.004	.010	.022	.041	.070	.109	.157	.209	.259	.296	.311	.294	.238	.149	.051
	7	.000	.000	.000	.000	.000	.000	.001	.003	.008	.016	.031	.055	.090	.137	.198	.267	.336	.385	.383	.279
	8	.000	.000	.000	.000	.000	.000	.000	.000	.001	.002	.004	.008	.017	.032	.058	.100	.168	.272	.430	.663
9	0	.914	.630	.387	.232	.134	.075	.040	.021	.010	.005	.002	.001	.000	.000	.000	.000	.000	.000	.000	.000
	1	.083	.299	.387	.368	.302	.225	.156	.100	.060	.034	.018	.008	.004	.001	.000	.000	.000	.000	.000	.000
	2	.003	.063	.172	.260	.302	.300	.267	.216	.161	.111	.070	.041	.021	.010	.004	.001	.000	.000	.000	.000
	3	.000	.008	.045	.107	.176	.234	.267	.272	.251	.212	.164	.116	.074	.042	.021	.009	.003	.001	.000	.000
	4	.000	.001	.007	.028	.066	.117	.172	.219	.251	.260	.246	.213	.167	.118	.074	.039	.017	.005	.001	.000
	5	.000	.000	.001	.005	.017	.039	.074	.118	.167	.213	.246	.260	.251	.219	.172	.117	.066	.028	.007	.001
	6	.000	.000	.000	.001	.003	.009	.021	.042	.074	.116	.164	.212	.251	.272	.267	.234	.176	.107	.045	.008
	7	.000	.000	.000	.000	.000	.001	.004	.010	.021	.041	.070	.111	.161	.216	.267	.300	.302	.260	.172	.063
	8	.000	.000	.000	.000	.000	.000	.000	.001	.004	.008	.018	.034	.060	.100	.156	.225	.302	.368	.387	.299
	9	.000	.000	.000	.000	.000	.000	.000	.000	.000	.001	.002	.005	.010	.021	.040	.075	.134	.232	.387	.630
10	0	.904	.599	.349	.197	.107	.056	.028	.014	.006	.003	.001	.000	.000	.000	.000	.000	.000	.000	.000	.000
	1	.091	.315	.387	.347	.268	.188	.121	.072	.040	.021	.010	.004	.002	.000	.000	.000	.000	.000	.000	.000
	2	.004	.075	.194	.276	.302	.282	.233	.176	.121	.076	.044	.023	.011	.004	.001	.000	.000	.000	.000	.000
	3	.000	.010	.057	.130	.201	.250	.267	.252	.215	.166	.117	.075	.042	.021	.009	.003	.001	.000	.000	.000
	4	.000	.001	.011	.040	.088	.146	.200	.238	.251	.238	.205	.160	.111	.069	.037	.016	.006	.001	.000	.000
	5	.000	.000	.001	.008	.026	.058	.103	.154	.201	.234	.246	.234	.201	.154	.103	.058	.026	.008	.001	.000
	6	.000	.000	.000	.001	.006	.016	.037	.069	.111	.160	.205	.238	.251	.238	.200	.146	.088	.040	.011	.001
	7	.000	.000	.000	.000	.001	.003	.009	.021	.042	.075	.117	.166	.215	.252	.267	.250	.201	.130	.057	.010
	8	.000	.000	.000	.000	.000	.000	.001	.004	.011	.023	.044	.076	.121	.176	.233	.282	.302	.276	.194	.075
	9	.000	.000	.000	.000	.000	.000	.000	.000	.002	.004	.010	.021	.040	.072	.121	.188	.268	.347	.387	.315
	10	.000	.000	.000	.000	.000	.000	.000	.000	.000	.001	.001	.003	.006	.014	.028	.056	.107	.197	.349	.599
11	0	.895	.569	.314	.167	.086	.042	.020	.009	.004	.001	.000	.000	.000	.000	.000	.000	.000	.000	.000	.000
	1	.099	.329	.384	.325	.236	.155	.093	.052	.027	.013	.005	.002	.001	.000	.000	.000	.000	.000	.000	.000
	2	.005	.087	.213	.287	.295	.258	.200	.140	.089	.051	.027	.013	.005	.002	.001	.000	.000	.000	.000	.000
	3	.000	.014	.071	.152	.221	.258	.257	.225	.177	.126	.081	.046	.023	.010	.004	.001	.000	.000	.000	.000
	4	.000	.001	.016	.054	.111	.172	.220	.243	.236	.206	.161	.113	.070	.038	.017	.006	.002	.000	.000	.000
	5	.000	.000	.002	.013	.039	.080	.132	.183	.221	.236	.226	.193	.147	.099	.057	.027	.010	.002	.000	.000

Table 4 continued

n	r	.01	.05	.10	.15	.20	.25	.30	.35	.40	.45	.50	.55	.60	.65	.70	.75	.80	.85	.90	.95
												p										
11	6	.000	.000	.000	.002	.010	.027	.057	.099	.147	.193	.226	.236	.221	.183	.132	.080	.039	.013	.002	.000
	7	.000	.000	.000	.000	.002	.006	.017	.038	.070	.113	.161	.206	.236	.243	.220	.172	.111	.054	.016	.001
	8	.000	.000	.000	.000	.000	.001	.004	.010	.023	.046	.081	.126	.177	.225	.257	.258	.221	.152	.071	.014
	9	.000	.000	.000	.000	.000	.000	.001	.002	.005	.013	.027	.051	.089	.140	.200	.258	.295	.287	.213	.087
	10	.000	.000	.000	.000	.000	.000	.000	.000	.001	.002	.005	.013	.027	.052	.093	.155	.236	.325	.384	.329
	11	.000	.000	.000	.000	.000	.000	.000	.000	.000	.000	.000	.001	.004	.009	.020	.042	.086	.167	.314	.569
12	0	.886	.540	.282	.142	.069	.032	.014	.006	.002	.001	.000	.000	.000	.000	.000	.000	.000	.000	.000	.000
	1	.107	.341	.377	.301	.206	.127	.071	.037	.017	.008	.003	.001	.000	.000	.000	.000	.000	.000	.000	.000
	2	.006	.099	.230	.292	.283	.232	.168	.109	.064	.034	.016	.007	.002	.001	.000	.000	.000	.000	.000	.000
	3	.000	.017	.085	.172	.236	.258	.240	.195	.142	.092	.054	.028	.012	.005	.001	.000	.000	.000	.000	.000
	4	.000	.002	.021	.068	.133	.194	.231	.237	.213	.170	.121	.076	.042	.020	.008	.002	.001	.000	.000	.000
	5	.000	.000	.004	.019	.053	.103	.158	.204	.227	.223	.193	.149	.101	.059	.029	.011	.003	.001	.000	.000
	6	.000	.000	.000	.004	.016	.040	.079	.128	.177	.212	.226	.212	.177	.128	.079	.040	.016	.004	.000	.000
	7	.000	.000	.000	.001	.003	.011	.029	.059	.101	.149	.193	.223	.227	.204	.158	.103	.053	.019	.004	.000
	8	.000	.000	.000	.000	.001	.002	.008	.020	.042	.076	.121	.170	.213	.237	.231	.194	.133	.068	.021	.002
	9	.000	.000	.000	.000	.000	.000	.001	.005	.012	.028	.054	.092	.142	.195	.240	.258	.236	.172	.085	.017
	10	.000	.000	.000	.000	.000	.000	.000	.001	.002	.007	.016	.034	.064	.109	.168	.232	.283	.292	.230	.099
	11	.000	.000	.000	.000	.000	.000	.000	.000	.000	.001	.003	.008	.017	.037	.071	.127	.206	.301	.377	.341
	12	.000	.000	.000	.000	.000	.000	.000	.000	.000	.000	.000	.001	.002	.006	.014	.032	.069	.142	.282	.540
15	0	.860	.463	.206	.087	.035	.013	.005	.002	.000	.000	.000	.000	.000	.000	.000	.000	.000	.000	.000	.000
	1	.130	.366	.343	.231	.132	.067	.031	.013	.005	.002	.000	.000	.000	.000	.000	.000	.000	.000	.000	.000
	2	.009	.135	.267	.286	.231	.156	.092	.048	.022	.009	.003	.001	.000	.000	.000	.000	.000	.000	.000	.000
	3	.000	.031	.129	.219	.250	.225	.170	.111	.063	.032	.014	.005	.002	.000	.000	.000	.000	.000	.000	.000
	4	.000	.005	.043	.116	.188	.225	.219	.179	.127	.078	.042	.019	.007	.002	.001	.000	.000	.000	.000	.000
	5	.000	.001	.010	.045	.103	.165	.206	.212	.186	.140	.092	.051	.024	.010	.003	.001	.000	.000	.000	.000
	6	.000	.000	.002	.013	.043	.092	.147	.191	.207	.191	.153	.105	.061	.030	.012	.003	.001	.000	.000	.000
	7	.000	.000	.000	.003	.014	.039	.081	.132	.177	.201	.196	.165	.118	.071	.035	.013	.003	.001	.000	.000
	8	.000	.000	.000	.001	.003	.013	.035	.071	.118	.165	.196	.201	.177	.132	.081	.039	.014	.003	.000	.000
	9	.000	.000	.000	.000	.001	.003	.012	.030	.061	.105	.153	.191	.207	.191	.147	.092	.043	.013	.002	.000
	10	.000	.000	.000	.000	.000	.001	.003	.010	.024	.051	.092	.140	.186	.212	.206	.165	.103	.045	.010	.001
	11	.000	.000	.000	.000	.000	.000	.001	.002	.007	.019	.042	.078	.127	.179	.219	.225	.188	.116	.043	.005
	12	.000	.000	.000	.000	.000	.000	.000	.000	.002	.005	.014	.032	.063	.111	.170	.225	.250	.219	.129	.031
	13	.000	.000	.000	.000	.000	.000	.000	.000	.000	.001	.003	.009	.022	.048	.092	.156	.231	.286	.267	.135
	14	.000	.000	.000	.000	.000	.000	.000	.000	.000	.000	.000	.002	.005	.013	.031	.067	.132	.231	.343	.366
	15	.000	.000	.000	.000	.000	.000	.000	.000	.000	.000	.000	.000	.000	.002	.005	.013	.035	.087	.206	.463
16	0	.851	.440	.185	.074	.028	.010	.003	.001	.000	.000	.000	.000	.000	.000	.000	.000	.000	.000	.000	.000
	1	.138	.371	.329	.210	.113	.053	.023	.009	.003	.001	.000	.000	.000	.000	.000	.000	.000	.000	.000	.000

Table 4 continued

n	r	.01	.05	.10	.15	.20	.25	.30	.35	.40	.45	.50	.55	.60	.65	.70	.75	.80	.85	.90	.95
16	2	.010	.146	.275	.277	.211	.134	.073	.035	.015	.006	.002	.001	.000	.000	.000	.000	.000	.000	.000	.000
	3	.000	.036	.142	.229	.246	.208	.146	.089	.047	.022	.009	.003	.001	.000	.000	.000	.000	.000	.000	.000
	4	.000	.006	.051	.131	.200	.225	.204	.155	.101	.057	.028	.011	.004	.001	.000	.000	.000	.000	.000	.000
	5	.000	.001	.014	.056	.120	.180	.210	.201	.162	.112	.067	.034	.014	.005	.001	.000	.000	.000	.000	.000
	6	.000	.000	.003	.018	.055	.110	.165	.198	.198	.168	.122	.075	.039	.017	.006	.001	.000	.000	.000	.000
	7	.000	.000	.000	.005	.020	.052	.101	.152	.189	.197	.175	.132	.084	.044	.019	.006	.001	.000	.000	.000
	8	.000	.000	.000	.001	.006	.020	.049	.092	.142	.181	.196	.181	.142	.092	.049	.020	.006	.001	.000	.000
	9	.000	.000	.000	.000	.001	.006	.019	.044	.084	.132	.175	.197	.189	.152	.101	.052	.020	.005	.000	.000
	10	.000	.000	.000	.000	.000	.001	.006	.017	.039	.075	.122	.168	.198	.198	.165	.110	.055	.018	.003	.000
	11	.000	.000	.000	.000	.000	.000	.001	.005	.014	.034	.067	.112	.162	.201	.210	.180	.120	.056	.014	.001
	12	.000	.000	.000	.000	.000	.000	.000	.001	.004	.011	.028	.057	.101	.155	.204	.225	.200	.131	.051	.006
	13	.000	.000	.000	.000	.000	.000	.000	.000	.001	.003	.009	.022	.047	.089	.146	.208	.246	.229	.142	.036
	14	.000	.000	.000	.000	.000	.000	.000	.000	.000	.001	.002	.006	.015	.035	.073	.134	.211	.277	.275	.146
	15	.000	.000	.000	.000	.000	.000	.000	.000	.000	.000	.000	.001	.003	.009	.023	.053	.113	.210	.329	.371
	16	.000	.000	.000	.000	.000	.000	.000	.000	.000	.000	.000	.000	.000	.001	.003	.010	.028	.074	.185	.440
20	0	.818	.358	.122	.039	.012	.003	.001	.000	.000	.000	.000	.000	.000	.000	.000	.000	.000	.000	.000	.000
	1	.165	.377	.270	.137	.058	.021	.007	.002	.000	.000	.000	.000	.000	.000	.000	.000	.000	.000	.000	.000
	2	.016	.189	.285	.229	.137	.067	.028	.010	.003	.001	.000	.000	.000	.000	.000	.000	.000	.000	.000	.000
	3	.001	.060	.190	.243	.205	.134	.072	.032	.012	.004	.001	.000	.000	.000	.000	.000	.000	.000	.000	.000
	4	.000	.013	.090	.182	.218	.190	.130	.074	.035	.014	.005	.001	.000	.000	.000	.000	.000	.000	.000	.000
	5	.000	.002	.032	.103	.175	.202	.179	.127	.075	.036	.015	.005	.001	.000	.000	.000	.000	.000	.000	.000
	6	.000	.000	.009	.045	.109	.169	.192	.171	.124	.075	.036	.015	.005	.001	.000	.000	.000	.000	.000	.000
	7	.000	.000	.002	.016	.055	.112	.164	.184	.166	.122	.074	.037	.015	.005	.001	.000	.000	.000	.000	.000
	8	.000	.000	.000	.005	.022	.061	.114	.161	.180	.162	.120	.073	.035	.014	.004	.001	.000	.000	.000	.000
	9	.000	.000	.000	.001	.007	.027	.065	.116	.160	.177	.160	.119	.071	.034	.012	.003	.000	.000	.000	.000
	10	.000	.000	.000	.000	.002	.010	.031	.069	.117	.159	.176	.159	.117	.069	.031	.010	.002	.000	.000	.000
	11	.000	.000	.000	.000	.000	.003	.012	.034	.071	.119	.160	.177	.160	.116	.065	.027	.007	.001	.000	.000
	12	.000	.000	.000	.000	.000	.001	.004	.014	.035	.073	.120	.162	.180	.161	.114	.061	.022	.005	.000	.000
	13	.000	.000	.000	.000	.000	.000	.001	.005	.015	.037	.074	.122	.166	.184	.164	.112	.055	.016	.002	.000
	14	.000	.000	.000	.000	.000	.000	.000	.001	.005	.015	.036	.075	.124	.171	.192	.169	.109	.045	.009	.000
	15	.000	.000	.000	.000	.000	.000	.000	.000	.001	.005	.015	.036	.075	.127	.179	.202	.175	.103	.032	.002
	16	.000	.000	.000	.000	.000	.000	.000	.000	.000	.001	.005	.014	.035	.074	.130	.190	.218	.182	.090	.013
	17	.000	.000	.000	.000	.000	.000	.000	.000	.000	.000	.001	.004	.012	.032	.072	.134	.205	.243	.190	.060
	18	.000	.000	.000	.000	.000	.000	.000	.000	.000	.000	.000	.001	.003	.010	.028	.067	.137	.229	.285	.189
	19	.000	.000	.000	.000	.000	.000	.000	.000	.000	.000	.000	.000	.000	.002	.007	.021	.058	.137	.270	.377
	20	.000	.000	.000	.000	.000	.000	.000	.000	.000	.000	.000	.000	.000	.000	.001	.003	.012	.039	.122	.358

p

Table 5 Poisson Probability Distribution

For a given value of λ, entry indicates the probability of obtaining a specified value of r.

r	λ .1	.2	.3	.4	.5	.6	.7	.8	.9	1.0
0	.9048	.8187	.7408	.6703	.6065	.5488	.4966	.4493	.4066	.3679
1	.0905	.1637	.2222	.2681	.3033	.3293	.3476	.3595	.3659	.3679
2	.0045	.0164	.0333	.0536	.0758	.0988	.1217	.1438	.1647	.1839
3	.0002	.0011	.0033	.0072	.0126	.0198	.0284	.0383	.0494	.0613
4	.0000	.0001	.0003	.0007	.0016	.0030	.0050	.0077	.0111	.0153
5	.0000	.0000	.0000	.0001	.0002	.0004	.0007	.0012	.0020	.0031
6	.0000	.0000	.0000	.0000	.0000	.0000	.0001	.0002	.0003	.0005
7	.0000	.0000	.0000	.0000	.0000	.0000	.0000	.0000	.0000	.0001

r	λ 1.1	1.2	1.3	1.4	1.5	1.6	1.7	1.8	1.9	2.0
0	.3329	.3012	.2725	.2466	.2231	.2019	.1827	.1653	.1496	.1353
1	.3662	.3614	.3543	.3452	.3347	.3230	.3106	.2975	.2842	.2707
2	.2014	.2169	.2303	.2417	.2510	.2584	.2640	.2678	.2700	.2707
3	.0738	.0867	.0998	.1128	.1255	.1378	.1496	.1607	.1710	.1804
4	.0203	.0260	.0324	.0395	.0471	.0551	.0636	.0723	.0812	.0902
5	.0045	.0062	.0084	.0111	.0141	.0176	.0216	.0260	.0309	.0361
6	.0008	.0012	.0018	.0026	.0035	.0047	.0061	.0078	.0098	.0120
7	.0001	.0002	.0003	.0005	.0008	.0011	.0015	.0020	.0027	.0034
8	.0000	.0000	.0001	.0001	.0001	.0002	.0003	.0005	.0006	.0009
9	.0000	.0000	.0000	.0000	.0000	.0000	.0001	.0001	.0001	.0002

r	λ 2.1	2.2	2.3	2.4	2.5	2.6	2.7	2.8	2.9	3.0
0	.1225	.1108	.1003	.0907	.0821	.0743	.0672	.0608	.0550	.0498
1	.2572	.2438	.2306	.2177	.2052	.1931	.1815	.1703	.1596	.1494
2	.2700	.2681	.2652	.2613	.2565	.2510	.2450	.2384	.2314	.2240
3	.1890	.1966	.2033	.2090	.2138	.2176	.2205	.2225	.2237	.2240
4	.0992	.1082	.1169	.1254	.1336	.1414	.1488	.1557	.1622	.1680
5	.0417	.0476	.0538	.0602	.0668	.0735	.0804	.0872	.0940	.1008
6	.0146	.0174	.0206	.0241	.0278	.0319	.0362	.0407	.0455	.0504
7	.0044	.0055	.0068	.0083	.0099	.0118	.0139	.0163	.0188	.0216
8	.0011	.0015	.0019	.0025	.0031	.0038	.0047	.0057	.0068	.0081
9	.0003	.0004	.0005	.0007	.0009	.0011	.0014	.0018	.0022	.0027
10	.0001	.0001	.0001	.0002	.0002	.0003	.0004	.0005	.0006	.0008
11	.0000	.0000	.0000	.0000	.0000	.0001	.0001	.0001	.0002	.0002
12	.0000	.0000	.0000	.0000	.0000	.0000	.0000	.0000	.0000	.0001

r	λ 3.1	3.2	3.3	3.4	3.5	3.6	3.7	3.8	3.9	4.0
0	.0450	.0408	.0369	.0334	.0302	.0273	.0247	.0224	.0202	.0183
1	.1397	.1304	.1217	.1135	.1057	.0984	.0915	.0850	.0789	.0733

Table 5 continued

r	3.1	3.2	3.3	3.4	λ 3.5	3.6	3.7	3.8	3.9	4.0
2	.2165	.2087	.2008	.1929	.1850	.1771	.1692	.1615	.1539	.1465
3	.2237	.2226	.2209	.2186	.2158	.2125	.2087	.2046	.2001	.1954
4	.1734	.1781	.1823	.1858	.1888	.1912	.1931	.1944	.1951	.1954
5	.1075	.1140	.1203	.1264	.1322	.1377	.1429	.1477	.1522	.1563
6	.0555	.0608	.0662	.0716	.0771	.0826	.0881	.0936	.0989	.1042
7	.0246	.2078	.0312	.0348	.0385	.0425	.0466	.0508	.0551	.0595
8	.0095	.0111	.0129	.0148	.0169	.0191	.0215	.0241	.0269	.0298
9	.0033	.0040	.0047	.0056	.0066	.0076	.0089	.0102	.0116	.0132
10	.0010	.0013	.0016	.0019	.0023	.0028	.0033	.0039	.0045	.0053
11	.0003	.0004	.0005	.0006	.0007	.0009	.0011	.0013	.0016	.0019
12	.0001	.0001	.0001	.0002	.0002	.0003	.0003	.0004	.0005	.0006
13	.0000	.0000	.0000	.0000	.0001	.0001	.0001	.0001	.0002	.0002
14	.0000	.0000	.0000	.0000	.0000	.0000	.0000	.0000	.0000	.0001

r	4.1	4.2	4.3	4.4	λ 4.5	4.6	4.7	4.8	4.9	5.0
0	.0166	.0150	.0136	.0123	.0111	.0101	.0091	.0082	.0074	.0067
1	.0679	.0630	.0583	.0540	.0500	.0462	.0427	.0395	.0365	.0337
2	.1393	.1323	.1254	.1188	.1125	.1063	.1005	.0948	.0894	.0842
3	.1904	.1852	.1798	.1743	.1687	.1631	.1574	.1517	.1460	.1404
4	.1951	.1944	.1933	.1917	.1898	.1875	.1849	.1820	.1789	.1755
5	.1600	.1633	.1662	.1687	.1708	.1725	.1738	.1747	.1753	.1755
6	.1093	.1143	.1191	.1237	.1281	.1323	.1362	.1398	.1432	.1462
7	.0640	.0686	.0732	.0778	.0824	.0869	.0914	.0959	.1002	.1044
8	.0328	.0360	.0393	.0428	.0463	.0500	.0537	.0575	.0614	.0653
9	.0150	.0168	.0188	.0209	.0232	.0255	.0280	.0307	.0334	.0363
10	.0061	.0071	.0081	.0092	.0104	.0118	.0132	.0147	.0164	.0181
11	.0023	.0027	.0032	.0037	.0043	.0049	.0056	.0064	.0073	.0082
12	.0008	.0009	.0011	.0014	.0016	.0019	.0022	.0026	.0030	.0034
13	.0002	.0003	.0004	.0005	.0006	.0007	.0008	.0009	.0011	.0013
14	.0001	.0001	.0001	.0001	.0002	.0002	.0003	.0003	.0004	.0005
15	.0000	.0000	.0000	.0000	.0001	.0001	.0001	.0001	.0001	.0002

r	5.1	5.2	5.3	5.4	λ 5.5	5.6	5.7	5.8	5.9	6.0
0	.0061	.0055	.0050	.0045	.0041	.0037	.0033	.0030	.0027	.0025
1	.0311	.0287	.0265	.0244	.0225	.0207	.0191	.0176	.0162	.0149
2	.0793	.0746	.0701	.0659	.0618	.0580	.0544	.0509	.0477	.0446
3	.1348	.1293	.1239	.1185	.1133	.1082	.1033	.0985	.0938	.0892
4	.1719	.1681	.1641	.1600	.1558	.1515	.1472	.1428	.1383	.1339
5	.1753	.1748	.1740	.1728	.1714	.1697	.1678	.1656	.1632	.1606
6	.1490	.1515	.1537	.1555	.1571	.1584	.1594	.1601	.1605	.1606
7	.1086	.1125	.1163	.1200	.1234	.1267	.1298	.1326	.1353	.1377

Table 5 continued

					λ					
r	5.1	5.2	5.3	5.4	5.5	5.6	5.7	5.8	5.9	6.0
8	.0692	.0731	.0771	.0810	.0849	.0887	.0925	.0962	.0998	.1033
9	.0392	.0423	.0454	.0486	.0519	.0552	.0586	.0620	.0654	.0688
10	.0200	.0220	.0241	.0262	.0285	.0309	.0334	.0359	.0386	.0413
11	.0093	.0104	.0116	.0129	.0143	.0157	.0173	.0190	.0207	.0225
12	.0039	.0045	.0051	.0058	.0065	.0073	.0082	.0092	.0102	.0113
13	.0015	.0018	.0021	.0024	.0028	.0032	.0036	.0041	.0046	.0052
14	.0006	.0007	.0008	.0009	.0011	.0013	.0015	.0017	.0019	.0022
15	.0002	.0002	.0003	.0003	.0004	.0005	.0006	.0007	.0008	.0009
16	.0001	.0001	.0001	.0001	.0001	.0002	.0002	.0002	.0003	.0003
17	.0000	.0000	.0000	.0000	.0000	.0000	.0001	.0001	.0001	.0001

					λ					
r	6.1	6.2	6.3	6.4	6.5	6.6	6.7	6.8	6.9	7.0
0	.0022	.0020	.0018	.0017	.0015	.0014	.0012	.0011	.0010	.0009
1	.0137	.0126	.0116	.0106	.0098	.0090	.0082	.0076	.0070	.0064
2	.0417	.0390	.0364	.0340	.0318	.0296	.0276	.0258	.0240	.0223
3	.0848	.0806	.0765	.0726	.0688	.0652	.0617	.0584	.0552	.0521
4	.1294	.1249	.1205	.1162	.1118	.1076	.1034	.0992	.0952	.0912
5	.1579	.1549	.1519	.1487	.1454	.1420	.1385	.1349	.1314	.1277
6	.1605	.1601	.1595	.1586	.1575	.1562	.1546	.1529	.1511	.1490
7	.1399	.1418	.1435	.1450	.1462	.1472	.1480	.1486	.1489	.1490
8	.1066	.1099	.1130	.1160	.1188	.1215	.1240	.1263	.1284	.1304
9	.0723	.0757	.0791	.0825	.0858	.0891	.0923	.0954	.0985	.1014
10	.0441	.0469	.0498	.0528	.0558	.0588	.0618	.0649	.0679	.0710
11	.0245	.0265	.0285	.0307	.0330	.0353	.0377	.0401	.0426	.0452
12	.0124	.0137	.0150	.0164	.0179	.0194	.0210	.0227	.0245	.0264
13	.0058	.0065	.0073	.0081	.0089	.0098	.0108	.0119	.0130	.0142
14	.0025	.0029	.0033	.0037	.0041	.0046	.0052	.0058	.0064	.0071
15	.0010	.0012	.0014	.0016	.0018	.0020	.0023	.0026	.0029	.0033
16	.0004	.0005	.0005	.0006	.0007	.0008	.0010	.0011	.0013	.0014
17	.0001	.0002	.0002	.0002	.0003	.0003	.0004	.0004	.0005	.0006
18	.0000	.0001	.0001	.0001	.0001	.0001	.0001	.0002	.0002	.0002
19	.0000	.0000	.0000	.0000	.0000	.0000	.0000	.0001	.0001	.0001

					λ					
r	7.1	7.2	7.3	7.4	7.5	7.6	7.7	7.8	7.9	8.0
0	.0008	.0007	.0007	.0006	.0006	.0005	.0005	.0004	.0004	.0003
1	.0059	.0054	.0049	.0045	.0041	.0038	.0035	.0032	.0029	.0027
2	.0208	.0194	.0180	.0167	.0156	.0145	.0134	.0125	.0116	.0107
3	.0492	.0464	.0438	.0413	.0389	.0366	.0345	.0324	.0305	.0286
4	.0874	.0836	.0799	.0764	.0729	.0696	.0663	.0632	.0602	.0573
5	.1241	.1204	.1167	.1130	.1094	.1057	.1021	.0986	.0951	.0916
6	.1468	.1445	.1420	.1394	.1367	.1339	.1311	.1282	.1252	.1221

Table 5 continued

					λ					
r	7.1	7.2	7.3	7.4	7.5	7.6	7.7	7.8	7.9	8.0
7	.1489	.1486	.1481	.1474	.1465	.1454	.1442	.1428	.1413	.1396
8	.1321	.1337	.1351	.1363	.1373	.1382	.1388	.1392	.1395	.1396
9	.1042	.1070	.1096	.1121	.1144	.1167	.1187	.1207	.1224	.1241
10	.0740	.0770	.0800	.0829	.0858	.0887	.0914	.0941	.0967	.0993
11	.0478	.0504	.0531	.0558	.0585	.0613	.0640	.0667	.0695	.0722
12	.0283	.0303	.0323	.0344	.0366	.0388	.0411	.0434	.0457	.0481
13	.0154	.0168	.0181	.0196	.0211	.0227	.0243	.0260	.0278	.0296
14	.0078	.0086	.0095	.0104	.0113	.0123	.0134	.0145	.0157	.0169
15	.0037	.0041	.0046	.0051	.0057	.0062	.0069	.0075	.0083	.0090
16	.0016	.0019	.0021	.0024	.0026	.0030	.0033	.0037	.0041	.0045
17	.0007	.0008	.0009	.0010	.0012	.0013	.0015	.0017	.0019	.0021
18	.0003	.0003	.0004	.0004	.0005	.0006	.0006	.0007	.0008	.0009
19	.0001	.0001	.0001	.0002	.0002	.0002	.0003	.0003	.0003	.0004
20	.0000	.0000	.0001	.0001	.0001	.0001	.0001	.0001	.0001	.0002
21	.0000	.0000	.0000	.0000	.0000	.0000	.0000	.0000	.0001	.0001

					λ					
r	8.1	8.2	8.3	8.4	8.5	8.6	8.7	8.8	8.9	9.0
0	.0003	.0003	.0002	.0002	.0002	.0002	.0002	.0002	.0001	.0001
1	.0025	.0023	.0021	.0019	.0017	.0016	.0014	.0013	.0012	.0011
2	.0100	.0092	.0086	.0079	.0074	.0068	.0063	.0058	.0054	.0050
3	.0269	.0252	.0237	.0222	.0208	.0195	.0183	.0171	.0160	.0150
4	.0544	.0517	.0491	.0466	.0443	.0420	.0398	.0377	.0357	.0337
5	.0882	.0849	.0816	.0784	.0752	.0722	.0692	.0663	.0635	.0607
6	.1191	.1160	.1128	.1097	.1066	.1034	.1003	.0972	.0941	.0911
7	.1378	.1358	.1338	.1317	.1294	.1271	.1247	.1222	.1197	.1171
8	.1395	.1392	.1388	.1382	.1375	.1366	.1356	.1344	.1332	.1318
9	.1256	.1269	.1280	.1290	.1299	.1306	.1311	.1315	.1317	.1318
10	.1017	.1040	.1063	.1084	.1104	.1123	.1140	.1157	.1172	.1186
11	.0749	.0776	.0802	.0828	.0853	.0878	.0902	.0925	.0948	.0970
12	.0505	.0530	.0555	.0579	.0604	.0629	.0654	.0679	.0703	.0728
13	.0315	.0334	.0354	.0374	.0395	.0416	.0438	.0459	.0481	.0504
14	.0182	.0196	.0210	.0225	.0240	.0256	.0272	.0289	.0306	.0324
15	.0098	.0107	.0116	.0126	.0136	.0147	.0158	.0169	.0182	.0194
16	.0050	.0055	.0060	.0066	.0072	.0079	.0086	.0093	.0101	.0109
17	.0024	.0026	.0029	.0033	.0036	.0040	.0044	.0048	.0053	.0058
18	.0011	.0012	.0014	.0015	.0017	.0019	.0021	.0024	.0026	.0029
19	.0005	.0005	.0006	.0007	.0008	.0009	.0010	.0011	.0012	.0014
20	.0002	.0002	.0002	.0003	.0003	.0004	.0004	.0005	.0005	.0006
21	.0001	.0001	.0001	.0001	.0001	.0002	.0002	.0002	.0002	.0003
22	.0000	.0000	.0000	.0000	.0001	.0001	.0001	.0001	.0001	.0001

Table 5 continued

					λ					
r	9.1	9.2	9.3	9.4	9.5	9.6	9.7	9.8	9.9	10
0	.0001	.0001	.0001	.0001	.0001	.0001	.0001	.0001	.0001	.0000
1	.0010	.0009	.0009	.0008	.0007	.0007	.0006	.0005	.0005	.0005
2	.0046	.0043	.0040	.0037	.0034	.0031	.0029	.0027	.0025	.0023
3	.0140	.0131	.0123	.0115	.0107	.0100	.0093	.0087	.0081	.0076
4	.0319	.0302	.0285	.0269	.0254	.0240	.0226	.0213	.0201	.0189
5	.0581	.0555	.0530	.0506	.0483	.0460	.0439	.0418	.0398	.0378
6	.0881	.0851	.0822	.0793	.0764	.0736	.0709	.0682	.0656	.0631
7	.1145	.1118	.1091	.1064	.1037	.1010	.0982	.0955	.0928	.0901
8	.1302	.1286	.1269	.1251	.1232	.1212	.1191	.1170	.1148	.1126
9	.1317	.1315	.1311	.1306	.1300	.1293	.1284	.1274	.1263	.1251
10	.1198	.1210	.1219	.1228	.1235	.1241	.1245	.1249	.1250	.1251
11	.0991	.1012	.1031	.1049	.1067	.1083	.1098	.1112	.1125	.1137
12	.0752	.0776	.0799	.0822	.0844	.0866	.0888	.0908	.0928	.0948
13	.0526	.0549	.0572	.0594	.0617	.0640	.0662	.0685	.0707	.0729
14	.0342	.0361	.0380	.0399	.0419	.0439	.0459	.0479	.0500	.0521
15	.0208	.0221	.0235	.0250	.0265	.0281	.0297	.0313	.0330	.0347
16	.0118	.0127	.0137	.0147	.0157	.0168	.0180	.0192	.0204	.0217
17	.0063	.0069	.0075	.0081	.0088	.0095	.0103	.0111	.0119	.0128
18	.0032	.0035	.0039	.0042	.0046	.0051	.0055	.0060	.0065	.0071
19	.0015	.0017	.0019	.0021	.0023	.0026	.0028	.0031	.0034	.0037
20	.0007	.0008	.0009	.0010	.0011	.0012	.0014	.0015	.0017	.0019
21	.0003	.0003	.0004	.0004	.0005	.0006	.0006	.0007	.0008	.0009
22	.0001	.0001	.0002	.0002	.0002	.0002	.0003	.0003	.0004	.0004
23	.0000	.0001	.0001	.0001	.0001	.0001	.0001	.0001	.0002	.0002
24	.0000	.0000	.0000	.0000	.0000	.0000	.0000	.0001	.0001	.0001

					λ					
r	11	12	13	14	15	16	17	18	19	20
0	.0000	.0000	.0000	.0000	.0000	.0000	.0000	.0000	.0000	.0000
1	.0002	.0001	.0000	.0000	.0000	.0000	.0000	.0000	.0000	.0000
2	.0010	.0004	.0002	.0001	.0000	.0000	.0000	.0000	.0000	.0000
3	.0037	.0018	.0008	.0004	.0002	.0001	.0000	.0000	.0000	.0000
4	.0102	.0053	.0027	.0013	.0006	.0003	.0001	.0001	.0000	.0000
5	.0224	.0127	.0070	.0037	.0019	.0010	.0005	.0002	.0001	.0001
6	.0411	.0255	.0152	.0087	.0048	.0026	.0014	.0007	.0004	.0002
7	.0646	.0437	.0281	.0174	.0104	.0060	.0034	.0018	.0010	.0005
8	.0888	.0655	.0457	.0304	.0194	.0120	.0072	.0042	.0024	.0013
9	.1085	.0874	.0661	.0473	.0324	.0213	.0135	.0083	.0050	.0029
10	.1194	.1048	.0859	.0663	.0486	.0341	.0230	.0150	.0095	.0058
11	.1194	.1144	.1015	.0844	.0663	.0496	.0355	.0245	.0164	.0106
12	.1094	.1144	.1099	.0984	.0829	.0661	.0504	.0368	.0259	.0176
13	.0926	.1056	.1099	.1060	.0956	.0814	.0658	.0509	.0378	.0271
14	.0728	.0905	.1021	.1060	.1024	.0930	.0800	.0655	.0514	.0387

Table 5 continued

| | | | | | λ | | | | | |
r	11	12	13	14	15	16	17	18	19	20
15	.0534	.0724	.0885	.0989	.1024	.0992	.0906	.0786	.0650	.0516
16	.0367	.0543	.0719	.0866	.0960	.0992	.0963	.0884	.0772	.0646
17	.0237	.0383	.0550	.0713	.0847	.0934	.0963	.0936	.0863	.0760
18	.0145	.0256	.0397	.0554	.0706	.0830	.0909	.0936	.0911	.0844
19	.0084	.0161	.0272	.0409	.0557	.0699	.0814	.0887	.0911	.0888
20	.0046	.0097	.0177	.0286	.0418	.0559	.0692	.0798	.0866	.0888
21	.0024	.0055	.0109	.0191	.0299	.0426	.0560	.0684	.0783	.0846
22	.0012	.0030	.0065	.0121	.0204	.0310	.0433	.0560	.0676	.0769
23	.0006	.0016	.0037	.0074	.0133	.0216	.0320	.0438	.0559	.0669
24	.0003	.0008	.0020	.0043	.0083	.0144	.0226	.0328	.0442	.0557
25	.0001	.0004	.0010	.0024	.0050	.0092	.0154	.0237	.0336	.0446
26	.0000	.0002	.0005	.0013	.0029	.0057	.0101	.0164	.0246	.0343
27	.0000	.0001	.0002	.0007	.0016	.0034	.0063	.0109	.0173	.0254
28	.0000	.0000	.0001	.0003	.0009	.0019	.0038	.0070	.0117	.0181
29	.0000	.0000	.0001	.0002	.0004	.0011	.0023	.0044	.0077	.0125
30	.0000	.0000	.0000	.0001	.0002	.0006	.0013	.0026	.0049	.0083
31	.0000	.0000	.0000	.0000	.0001	.0003	.0007	.0015	.0030	.0054
32	.0000	.0000	.0000	.0000	.0001	.0001	.0004	.0009	.0018	.0034
33	.0000	.0000	.0000	.0000	.0000	.0001	.0002	.0005	.0010	.0020
34	.0000	.0000	.0000	.0000	.0000	.0000	.0001	.0002	.0006	.0012
35	.0000	.0000	.0000	.0000	.0000	.0000	.0000	.0001	.0003	.0007
36	.0000	.0000	.0000	.0000	.0000	.0000	.0000	.0001	.0002	.0004
37	.0000	.0000	.0000	.0000	.0000	.0000	.0000	.0000	.0001	.0002
38	.0000	.0000	.0000	.0000	.0000	.0000	.0000	.0000	.0000	.0001
39	.0000	.0000	.0000	.0000	.0000	.0000	.0000	.0000	.0000	.0001

Source: Extracted from William H. Beyer, ed., *CRC Basic Statistical Tables* (Cleveland, Ohio: The Chemical Rubber Co., 1971).

Table 6 Areas of a Standard Normal Distribution

The table entries represent the area under the standard normal curve from 0 to the specified value of z.

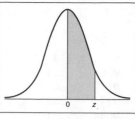

z	.00	.01	.02	.03	.04	.05	.06	.07	.08	.09
0.0	.0000	.0040	.0080	.0120	.0160	.0199	.0239	.0279	.0319	.0359
0.1	.0398	.0438	.0478	.0517	.0557	.0596	.0636	.0675	.0714	.0753
0.2	.0793	.0832	.0871	.0910	.0948	.0987	.1026	.1064	.1103	.1141
0.3	.1179	.1217	.1255	.1293	.1331	.1368	.1406	.1443	.1480	.1517
0.4	.1554	.1591	.1628	.1664	.1700	.1736	.1772	.1808	.1844	.1879
0.5	.1915	.1950	.1985	.2019	.2054	.2088	.2123	.2157	.2190	.2224
0.6	.2257	.2291	.2324	.2357	.2389	.2422	.2454	.2486	.2517	.2549
0.7	.2580	.2611	.2642	.2673	.2704	.2734	.2764	.2794	.2823	.2852
0.8	.2881	.2910	.2939	.2967	.2995	.3023	.3051	.3078	.3106	.3133
0.9	.3159	.3186	.3212	.3238	.3264	.3289	.3315	.3340	.3365	.3389
1.0	.3413	.3438	.3461	.3485	.3508	.3531	.3554	.3577	.3599	.3621
1.1	.3643	.3665	.3686	.3708	.3729	.3749	.3770	.3790	.3810	.3830
1.2	.3849	.3869	.3888	.3907	.3925	.3944	.3962	.3980	.3997	.4015
1.3	.4032	.4049	.4066	.4082	.4099	.4115	.4131	.4147	.4162	.4177
1.4	.4192	.4207	.4222	.4236	.4251	.4265	.4279	.4292	.4306	.4319
1.5	.4332	.4345	.4357	.4370	.4382	.4394	.4406	.4418	.4429	.4441
1.6	.4452	.4463	.4474	.4484	.4495	.4505	.4515	.4525	.4535	.4545
1.7	.4554	.4564	.4573	.4582	.4591	.4599	.4608	.4616	.4625	.4633
1.8	.4641	.4649	.4656	.4664	.4671	.4678	.4686	.4693	.4699	.4706
1.9	.4713	.4719	.4726	.4732	.4738	.4744	.4750	.4756	.4761	.4767
2.0	.4772	.4778	.4783	.4788	.4793	.4798	.4803	.4808	.4812	.4817
2.1	.4821	.4826	.4830	.4834	.4838	.4842	.4846	.4850	.4854	.4857
2.2	.4861	.4864	.4868	.4871	.4875	.4878	.4881	.4884	.4887	.4890
2.3	.4893	.4896	.4898	.4901	.4904	.4906	.4909	.4911	.4913	.4916
2.4	.4918	.4920	.4922	.4925	.4927	.4929	.4931	.4932	.4934	.4936
2.5	.4938	.4940	.4941	.4943	.4945	.4946	.4948	.4949	.4951	.4952
2.6	.4953	.4955	.4956	.4957	.4959	.4960	.4961	.4962	.4963	.4964
2.7	.4965	.4966	.4967	.4968	.4969	.4970	.4971	.4972	.4973	.4974
2.8	.4974	.4975	.4976	.4977	.4977	.4978	.4979	.4979	.4980	.4981
2.9	.4981	.4982	.4982	.4983	.4984	.4984	.4985	.4985	.4986	.4986
3.0	.4987	.4987	.4987	.4988	.4988	.4989	.4989	.4989	.4990	.4990
3.1	.4990	.4991	.4991	.4991	.4992	.4992	.4992	.4992	.4993	.4993
3.2	.4993	.4993	.4994	.4994	.4994	.4994	.4994	.4995	.4995	.4995
3.3	.4995	.4995	.4995	.4996	.4996	.4996	.4996	.4996	.4996	.4997
3.4	.4997	.4997	.4997	.4997	.4997	.4997	.4997	.4997	.4997	.4998
3.5	.4998	.4998	.4998	.4998	.4998	.4998	.4998	.4998	.4998	.4998
3.6	.4998	.4998	.4998	.4999	.4999	.4999	.4999	.4999	.4999	.4999

For values of z greater than or equal 3.70, use 0.4999 to approximate the shaded area under the standard normal curve.

Table 7 Student's t Distribution

Student's t values generated by Minitab Version 9.2

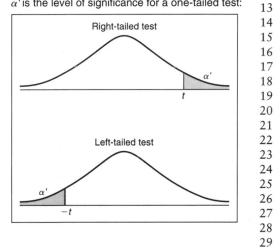

c is a confidence level:

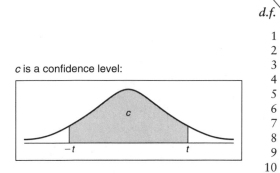

α′ is the level of significance for a one-tailed test:

Right-tailed test

Left-tailed test

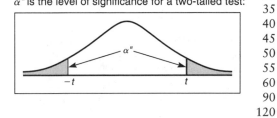

α″ is the level of significance for a two-tailed test:

d.f.	c 0.750 α' 0.125 α'' 0.250	0.800 0.100 0.200	0.850 0.075 0.150	0.900 0.050 0.100	0.950 0.025 0.050	0.980 0.010 0.020	0.990 0.005 0.010
1	2.414	3.078	4.165	6.314	12.706	31.821	63.657
2	1.604	1.886	2.282	2.920	4.303	6.965	9.925
3	1.423	1.638	1.924	2.353	3.182	4.541	5.841
4	1.344	1.533	1.778	2.132	2.776	3.747	4.604
5	1.301	1.476	1.699	2.015	2.571	3.365	4.032
6	1.273	1.440	1.650	1.943	2.447	3.143	3.707
7	1.254	1.415	1.617	1.895	2.365	2.998	3.499
8	1.240	1.397	1.592	1.860	2.306	2.896	3.355
9	1.230	1.383	1.574	1.833	2.262	2.821	3.250
10	1.221	1.372	1.559	1.812	2.228	2.764	3.169
11	1.214	1.363	1.548	1.796	2.201	2.718	3.106
12	1.209	1.356	1.538	1.782	2.179	2.681	3.055
13	1.204	1.350	1.530	1.771	2.160	2.650	3.012
14	1.200	1.345	1.523	1.761	2.145	2.624	2.977
15	1.197	1.341	1.517	1.753	2.131	2.602	2.947
16	1.194	1.337	1.512	1.746	2.120	2.583	2.921
17	1.191	1.333	1.508	1.740	2.110	2.567	2.898
18	1.189	1.330	1.504	1.734	2.101	2.552	2.878
19	1.187	1.328	1.500	1.729	2.093	2.539	2.861
20	1.185	1.325	1.497	1.725	2.086	2.528	2.845
21	1.183	1.323	1.494	1.721	2.080	2.518	2.831
22	1.182	1.321	1.492	1.717	2.074	2.508	2.819
23	1.180	1.319	1.489	1.714	2.069	2.500	2.807
24	1.179	1.318	1.487	1.711	2.064	2.492	2.797
25	1.178	1.316	1.485	1.708	2.060	2.485	2.787
26	1.177	1.315	1.483	1.706	2.056	2.479	2.779
27	1.176	1.314	1.482	1.703	2.052	2.473	2.771
28	1.175	1.313	1.480	1.701	2.048	2.467	2.763
29	1.174	1.311	1.479	1.699	2.045	2.462	2.756
30	1.173	1.310	1.477	1.697	2.042	2.457	2.750
35	1.170	1.306	1.472	1.690	2.030	2.438	2.724
40	1.167	1.303	1.468	1.684	2.021	2.423	2.704
45	1.165	1.301	1.465	1.679	2.014	2.412	2.690
50	1.164	1.299	1.462	1.676	2.009	2.403	2.678
55	1.163	1.297	1.460	1.673	2.004	2.396	2.668
60	1.162	1.296	1.458	1.671	2.000	2.390	2.660
90	1.158	1.291	1.452	1.662	1.987	2.369	2.632
120	1.156	1.289	1.449	1.658	1.980	2.358	2.617
∞	1.150	1.282	1.440	1.645	1.960	2.326	2.58

Table 8 Critical Values of Pearson Product-Moment Correlation, *r*

For a right-tailed test, use a positive *r* value:

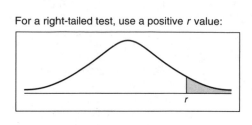

For a left-tailed test, use a negative *r* value:

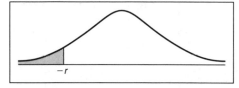

For a two-tailed test, use a positive *r* value and negative *r* value:

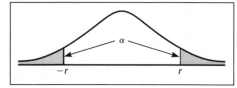

n	$\alpha = 0.01$ one tail	$\alpha = 0.01$ two tails	$\alpha = 0.05$ one tail	$\alpha = 0.05$ two tails
3	1.00	1.00	.99	1.00
4	.98	.99	.90	.95
5	.93	.96	.81	.88
6	.88	.92	.73	.81
7	.83	.87	.67	.75
8	.79	.83	.62	.71
9	.75	.80	.58	.67
10	.72	.76	.54	.63
11	.69	.73	.52	.60
12	.66	.71	.50	.58
13	.63	.68	.48	.53
14	.61	.66	.46	.53
15	.59	.64	.44	.51
16	.57	.61	.42	.50
17	.56	.61	.41	.48
18	.54	.59	.40	.47
19	.53	.58	.39	.46
20	.52	.56	.38	.44
21	.50	.55	.37	.43
22	.49	.54	.36	.42
23	.48	.53	.35	.41
24	.47	.52	.34	.40
25	.46	.51	.34	.40
26	.45	.50	.33	.39
27	.45	.49	.32	.38
28	.44	.48	.32	.37
29	.43	.47	.31	.37
30	.42	.46	.31	.36

Table 9 The χ^2 Distribution

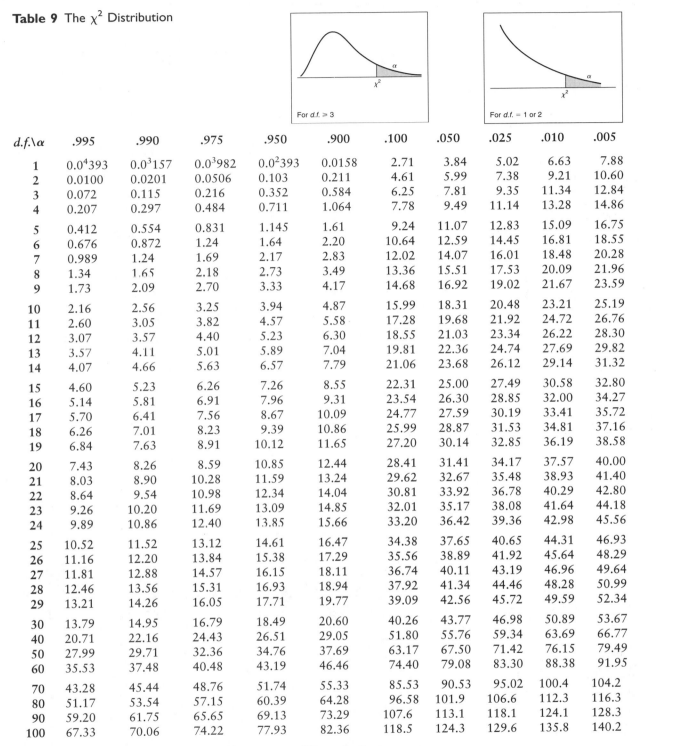

For d.f. ≥ 3

For d.f. = 1 or 2

d.f.\α	.995	.990	.975	.950	.900	.100	.050	.025	.010	.005
1	0.0^4393	0.0^3157	0.0^3982	0.0^2393	0.0158	2.71	3.84	5.02	6.63	7.88
2	0.0100	0.0201	0.0506	0.103	0.211	4.61	5.99	7.38	9.21	10.60
3	0.072	0.115	0.216	0.352	0.584	6.25	7.81	9.35	11.34	12.84
4	0.207	0.297	0.484	0.711	1.064	7.78	9.49	11.14	13.28	14.86
5	0.412	0.554	0.831	1.145	1.61	9.24	11.07	12.83	15.09	16.75
6	0.676	0.872	1.24	1.64	2.20	10.64	12.59	14.45	16.81	18.55
7	0.989	1.24	1.69	2.17	2.83	12.02	14.07	16.01	18.48	20.28
8	1.34	1.65	2.18	2.73	3.49	13.36	15.51	17.53	20.09	21.96
9	1.73	2.09	2.70	3.33	4.17	14.68	16.92	19.02	21.67	23.59
10	2.16	2.56	3.25	3.94	4.87	15.99	18.31	20.48	23.21	25.19
11	2.60	3.05	3.82	4.57	5.58	17.28	19.68	21.92	24.72	26.76
12	3.07	3.57	4.40	5.23	6.30	18.55	21.03	23.34	26.22	28.30
13	3.57	4.11	5.01	5.89	7.04	19.81	22.36	24.74	27.69	29.82
14	4.07	4.66	5.63	6.57	7.79	21.06	23.68	26.12	29.14	31.32
15	4.60	5.23	6.26	7.26	8.55	22.31	25.00	27.49	30.58	32.80
16	5.14	5.81	6.91	7.96	9.31	23.54	26.30	28.85	32.00	34.27
17	5.70	6.41	7.56	8.67	10.09	24.77	27.59	30.19	33.41	35.72
18	6.26	7.01	8.23	9.39	10.86	25.99	28.87	31.53	34.81	37.16
19	6.84	7.63	8.91	10.12	11.65	27.20	30.14	32.85	36.19	38.58
20	7.43	8.26	8.59	10.85	12.44	28.41	31.41	34.17	37.57	40.00
21	8.03	8.90	10.28	11.59	13.24	29.62	32.67	35.48	38.93	41.40
22	8.64	9.54	10.98	12.34	14.04	30.81	33.92	36.78	40.29	42.80
23	9.26	10.20	11.69	13.09	14.85	32.01	35.17	38.08	41.64	44.18
24	9.89	10.86	12.40	13.85	15.66	33.20	36.42	39.36	42.98	45.56
25	10.52	11.52	13.12	14.61	16.47	34.38	37.65	40.65	44.31	46.93
26	11.16	12.20	13.84	15.38	17.29	35.56	38.89	41.92	45.64	48.29
27	11.81	12.88	14.57	16.15	18.11	36.74	40.11	43.19	46.96	49.64
28	12.46	13.56	15.31	16.93	18.94	37.92	41.34	44.46	48.28	50.99
29	13.21	14.26	16.05	17.71	19.77	39.09	42.56	45.72	49.59	52.34
30	13.79	14.95	16.79	18.49	20.60	40.26	43.77	46.98	50.89	53.67
40	20.71	22.16	24.43	26.51	29.05	51.80	55.76	59.34	63.69	66.77
50	27.99	29.71	32.36	34.76	37.69	63.17	67.50	71.42	76.15	79.49
60	35.53	37.48	40.48	43.19	46.46	74.40	79.08	83.30	88.38	91.95
70	43.28	45.44	48.76	51.74	55.33	85.53	90.53	95.02	100.4	104.2
80	51.17	53.54	57.15	60.39	64.28	96.58	101.9	106.6	112.3	116.3
90	59.20	61.75	65.65	69.13	73.29	107.6	113.1	118.1	124.1	128.3
100	67.33	70.06	74.22	77.93	82.36	118.5	124.3	129.6	135.8	140.2

From H. L. Herter, *Biometrika*, June 1964. Printed by permission of Biometrika Trustees.

Table 10 The F Distribution, 5% (Roman Type) and 1% (Boldface Type) Points for the Distribution of F*

Degrees of Freedom for Numerator

Degrees of Freedom for Denominator

	1	2	3	4	5	6	7	8	9	10	11	12	14	16	20	24	30	40	50	75	100	200	500	∞
1	161	200	216	225	230	234	237	239	241	242	243	244	245	246	248	249	250	251	252	253	253	254	254	254
	4052	**4999**	**5403**	**5625**	**5764**	**5859**	**5928**	**5981**	**6022**	**6056**	**6082**	**6106**	**6142**	**6169**	**6208**	**6234**	**6258**	**6286**	**6302**	**6323**	**6334**	**6352**	**6361**	**6366**
2	18.51	19.00	19.16	19.25	19.30	19.33	19.36	19.37	19.38	19.39	19.40	19.41	19.42	19.43	19.44	19.45	19.46	19.47	19.47	19.48	19.49	19.49	19.50	19.50
	98.49	**99.01**	**99.17**	**99.25**	**99.30**	**99.33**	**99.34**	**99.36**	**99.38**	**99.40**	**99.41**	**99.42**	**99.43**	**99.44**	**99.45**	**99.46**	**99.47**	**99.48**	**99.48**	**99.49**	**99.49**	**99.49**	**99.50**	**99.50**
3	10.13	9.55	9.28	9.12	9.01	8.94	8.88	8.84	8.81	8.78	8.76	8.74	8.71	8.69	8.66	8.64	8.62	8.60	8.58	8.57	8.56	8.54	8.54	8.53
	34.12	**30.81**	**29.46**	**28.71**	**28.24**	**27.91**	**27.67**	**27.49**	**27.34**	**27.23**	**27.13**	**27.05**	**26.92**	**26.83**	**26.69**	**26.60**	**26.50**	**26.41**	**26.30**	**26.27**	**26.23**	**26.18**	**26.14**	**26.12**
4	7.71	6.94	6.59	6.39	6.26	6.16	6.09	6.04	6.00	5.96	5.93	5.91	5.87	5.84	5.80	5.77	5.74	5.71	5.70	5.68	5.66	5.65	5.64	5.63
	21.20	**18.00**	**16.69**	**15.98**	**15.52**	**15.21**	**14.98**	**14.80**	**14.66**	**14.54**	**14.45**	**14.37**	**14.24**	**14.15**	**14.02**	**13.93**	**13.83**	**13.74**	**13.69**	**13.61**	**13.57**	**13.52**	**13.48**	**13.46**
5	6.61	5.79	5.41	5.19	5.05	4.95	4.88	4.82	4.78	4.74	4.70	4.68	4.64	4.60	4.56	4.53	4.50	4.46	4.44	4.42	4.40	4.38	4.37	4.36
	16.26	**13.27**	**12.06**	**11.39**	**10.97**	**10.67**	**10.45**	**10.27**	**10.15**	**10.05**	**9.96**	**9.89**	**9.77**	**9.68**	**9.55**	**9.47**	**9.38**	**9.29**	**9.24**	**9.17**	**9.13**	**9.07**	**9.04**	**9.02**
6	5.99	5.14	4.76	4.53	4.39	4.28	4.21	4.15	4.10	4.06	4.03	4.00	3.96	3.92	3.87	3.84	3.81	3.77	3.75	3.72	3.71	3.69	3.68	3.67
	13.74	**10.92**	**9.78**	**9.15**	**8.75**	**8.47**	**8.26**	**8.10**	**7.98**	**7.87**	**7.79**	**7.72**	**7.60**	**7.52**	**7.39**	**7.31**	**7.23**	**7.14**	**7.09**	**7.02**	**6.99**	**6.94**	**6.90**	**6.88**
7	5.59	4.74	4.35	4.12	3.97	3.87	3.79	3.73	3.68	3.63	3.60	3.57	3.52	3.49	3.44	3.41	3.38	3.34	3.32	3.29	3.28	3.25	3.24	3.23
	12.25	**9.55**	**8.45**	**7.85**	**7.46**	**7.19**	**7.00**	**6.84**	**6.71**	**6.62**	**6.54**	**6.47**	**6.35**	**6.27**	**6.15**	**6.07**	**5.98**	**5.90**	**5.85**	**5.78**	**5.75**	**5.70**	**5.67**	**5.65**
8	5.32	4.46	4.07	3.84	3.69	3.58	3.50	3.44	3.39	3.34	3.31	3.28	3.23	3.20	3.15	3.12	3.08	3.05	3.03	3.00	2.98	2.96	2.94	2.93
	11.26	**8.65**	**7.59**	**7.01**	**6.63**	**6.37**	**6.19**	**6.03**	**5.91**	**5.82**	**5.74**	**5.67**	**5.56**	**5.48**	**5.36**	**5.28**	**5.20**	**5.11**	**5.06**	**5.00**	**4.96**	**4.91**	**4.88**	**4.86**
9	5.12	4.26	3.86	3.63	3.48	3.37	3.29	3.23	3.18	3.13	3.10	3.07	3.02	2.98	2.93	2.90	2.86	2.82	2.80	2.77	2.76	2.73	2.72	2.71
	10.56	**8.02**	**6.99**	**6.42**	**6.06**	**5.80**	**5.62**	**5.47**	**5.35**	**5.26**	**5.18**	**5.11**	**5.00**	**4.92**	**4.80**	**4.73**	**4.64**	**4.56**	**4.51**	**4.45**	**4.41**	**4.36**	**4.33**	**4.31**
10	4.96	4.10	3.71	3.48	3.33	3.22	3.14	3.07	3.02	2.97	2.94	2.91	2.86	2.82	2.77	2.74	2.70	2.67	2.64	2.61	2.59	2.56	2.55	2.54
	10.04	**7.56**	**6.55**	**5.99**	**5.64**	**5.39**	**5.21**	**5.06**	**4.95**	**4.85**	**4.78**	**4.71**	**4.60**	**4.52**	**4.41**	**4.33**	**4.25**	**4.17**	**4.12**	**4.05**	**4.01**	**3.96**	**3.93**	**3.91**
11	4.84	3.98	3.59	3.36	3.20	3.09	3.01	2.95	2.90	2.86	2.82	2.79	2.74	2.70	2.65	2.61	2.57	2.53	2.50	2.47	2.45	2.42	2.41	2.40
	9.65	**7.20**	**6.22**	**5.67**	**5.32**	**5.07**	**4.88**	**4.74**	**4.63**	**4.54**	**4.46**	**4.40**	**4.29**	**4.21**	**4.10**	**4.02**	**3.94**	**3.86**	**3.80**	**3.74**	**3.70**	**3.66**	**3.62**	**3.60**
12	4.75	3.88	3.49	3.26	3.11	3.00	2.92	2.85	2.80	2.76	2.72	2.69	2.64	2.60	2.54	2.50	2.46	2.42	2.40	2.36	2.35	2.32	2.31	2.30
	9.33	**6.93**	**5.95**	**5.41**	**5.06**	**4.82**	**4.65**	**4.50**	**4.39**	**4.30**	**4.22**	**4.16**	**4.05**	**3.98**	**3.86**	**3.78**	**3.70**	**3.61**	**3.56**	**3.49**	**3.46**	**3.41**	**3.38**	**3.36**
13	4.67	3.80	3.41	3.18	3.02	2.92	2.84	2.77	2.72	2.67	2.63	2.60	2.55	2.51	2.46	2.42	2.38	2.34	2.32	2.28	2.26	2.24	2.22	2.21
	9.07	**6.70**	**5.74**	**5.20**	**4.86**	**4.62**	**4.44**	**4.30**	**4.19**	**4.10**	**4.02**	**3.96**	**3.85**	**3.78**	**3.67**	**3.59**	**3.51**	**3.42**	**3.37**	**3.30**	**3.27**	**3.21**	**3.18**	**3.16**
14	4.60	3.74	3.34	3.11	2.96	2.85	2.77	2.70	2.65	2.60	2.56	2.53	2.48	2.44	2.39	2.35	2.31	2.27	2.24	2.21	2.19	2.16	2.14	2.13
	8.86	**6.51**	**5.56**	**5.03**	**4.69**	**4.46**	**4.28**	**4.14**	**4.03**	**3.94**	**3.86**	**3.80**	**3.70**	**3.62**	**3.51**	**3.43**	**3.34**	**3.26**	**3.21**	**3.14**	**3.11**	**3.06**	**3.02**	**3.00**
15	4.54	3.68	3.29	3.06	2.90	2.79	2.70	2.64	2.59	2.55	2.51	2.48	2.43	2.39	2.33	2.29	2.25	2.21	2.18	2.15	2.12	2.10	2.08	2.07
	8.68	**6.36**	**5.42**	**4.89**	**4.56**	**4.32**	**4.14**	**4.00**	**3.89**	**3.80**	**3.73**	**3.67**	**3.56**	**3.48**	**3.36**	**3.29**	**3.20**	**3.12**	**3.07**	**3.00**	**2.97**	**2.92**	**2.89**	**2.87**
16	4.49	3.63	3.24	3.01	2.85	2.74	2.66	2.59	2.54	2.49	2.45	2.42	2.37	2.33	2.28	2.24	2.20	2.16	2.13	2.09	2.07	2.04	2.02	2.01
	8.53	**6.23**	**5.29**	**4.77**	**4.44**	**4.20**	**4.03**	**3.89**	**3.78**	**3.69**	**3.61**	**3.55**	**3.45**	**3.37**	**3.25**	**3.18**	**3.10**	**3.01**	**2.96**	**2.89**	**2.86**	**2.80**	**2.77**	**2.75**
17	4.45	3.59	3.20	2.96	2.81	2.70	2.62	2.55	2.50	2.45	2.41	2.38	2.33	2.29	2.23	2.19	2.15	2.11	2.08	2.04	2.02	1.99	1.97	1.96
	8.40	**6.11**	**5.18**	**4.67**	**4.34**	**4.10**	**3.93**	**3.79**	**3.68**	**3.59**	**3.52**	**3.45**	**3.35**	**3.27**	**3.16**	**3.08**	**3.00**	**2.92**	**2.86**	**2.79**	**2.76**	**2.70**	**2.67**	**2.65**

Table 10 continued

Degrees of Freedom for Numerator

Degrees of Freedom for Denominator	1	2	3	4	5	6	7	8	9	10	11	12	14	16	20	24	30	40	50	75	100	200	500	∞
18	4.41 8.28	3.55 6.01	3.16 5.09	2.93 4.58	2.77 4.25	2.66 4.01	2.58 3.85	2.51 3.71	2.46 3.60	2.41 3.51	2.37 3.44	2.34 3.37	2.29 3.27	2.25 3.19	2.19 3.07	2.15 3.00	2.11 2.91	2.07 2.83	2.04 2.78	2.00 2.71	1.98 2.68	1.95 2.62	1.93 2.59	1.92 2.57
19	4.38 8.18	3.52 5.93	3.13 5.01	2.90 4.50	2.74 4.17	2.63 3.94	2.55 3.77	2.48 3.63	2.43 3.52	2.38 3.43	2.34 3.36	2.31 3.30	2.26 3.19	2.21 3.12	2.15 3.00	2.11 2.92	2.07 2.84	2.02 2.76	2.00 2.70	1.96 2.63	1.94 2.60	1.91 2.54	1.90 2.51	1.88 2.49
20	4.35 8.10	3.49 5.85	3.10 4.94	2.87 4.43	2.71 4.10	2.60 3.87	2.52 3.71	2.45 3.56	2.40 3.45	2.35 3.37	2.31 3.30	2.28 3.23	2.23 3.13	2.18 3.05	2.12 2.94	2.08 2.86	2.04 2.77	1.99 2.69	1.96 2.63	1.92 2.56	1.90 2.53	1.87 2.47	1.85 2.44	1.84 2.42
21	4.32 8.02	3.47 5.78	3.07 4.87	2.84 4.37	2.68 4.04	2.57 3.81	2.49 3.65	2.42 3.51	2.37 3.40	2.32 3.31	2.28 3.24	2.25 3.17	2.20 3.07	2.15 2.99	2.09 2.88	2.05 2.80	2.00 2.72	1.96 2.63	1.93 2.58	1.89 2.51	1.87 2.47	1.84 2.42	1.82 2.38	1.81 2.36
22	4.30 7.94	3.44 5.72	3.05 4.82	2.82 4.31	2.66 3.99	2.55 3.76	2.47 3.59	2.40 3.45	2.35 3.35	2.30 3.26	2.26 3.18	2.23 3.12	2.18 3.02	2.13 2.94	2.07 2.83	2.03 2.75	1.98 2.67	1.93 2.58	1.91 2.53	1.87 2.46	1.84 2.42	1.81 2.37	1.80 2.33	1.78 2.31
23	4.28 7.88	3.42 5.66	3.03 4.76	2.80 4.26	2.64 3.94	2.53 3.71	2.45 3.54	2.38 3.41	2.32 3.30	2.28 3.21	2.24 3.14	2.20 3.07	2.14 2.97	2.10 2.89	2.04 2.78	2.00 2.70	1.96 2.62	1.91 2.53	1.88 2.48	1.84 2.41	1.82 2.37	1.79 2.32	1.77 2.28	1.76 2.26
24	4.26 7.82	3.40 5.61	3.01 4.72	2.78 4.22	2.62 3.90	2.51 3.67	2.43 3.50	2.36 3.36	2.30 3.25	2.26 3.17	2.22 3.09	2.18 3.03	2.13 2.93	2.09 2.85	2.02 2.74	1.98 2.66	1.94 2.58	1.89 2.49	1.86 2.44	1.82 2.36	1.80 2.33	1.76 2.27	1.74 2.23	1.73 2.21
25	4.24 7.77	3.38 5.57	2.99 4.68	2.76 4.18	2.60 3.86	2.49 3.63	2.41 3.46	2.34 3.32	2.28 3.21	2.24 3.13	2.20 3.05	2.16 2.99	2.11 2.89	2.06 2.81	2.00 2.70	1.96 2.62	1.92 2.54	1.87 2.45	1.84 2.40	1.80 2.32	1.77 2.29	1.74 2.23	1.72 2.19	1.71 2.17
26	4.22 7.72	3.37 5.53	2.98 4.64	2.74 4.14	2.59 3.82	2.47 3.59	2.39 3.42	2.32 3.29	2.27 3.17	2.22 3.09	2.18 3.02	2.15 2.96	2.10 2.86	2.05 2.77	1.99 2.66	1.95 2.58	1.90 2.50	1.85 2.41	1.82 2.36	1.78 2.28	1.76 2.25	1.72 2.19	1.70 2.15	1.69 2.13
27	4.21 7.68	3.35 5.49	2.96 4.60	2.73 4.11	2.57 3.79	2.46 3.56	2.37 3.39	2.30 3.26	2.25 3.14	2.20 3.06	2.16 2.98	2.13 2.93	2.08 2.83	2.03 2.74	1.97 2.63	1.93 2.55	1.88 2.47	1.84 2.38	1.80 2.33	1.76 2.25	1.74 2.21	1.71 2.16	1.68 2.12	1.67 2.10
28	4.20 7.64	3.34 5.45	2.95 4.57	2.71 4.07	2.56 3.76	2.44 3.53	2.36 3.36	2.29 3.23	2.24 3.11	2.19 3.03	2.15 2.95	2.12 2.90	2.06 2.80	2.02 2.71	1.96 2.60	1.91 2.52	1.87 2.44	1.81 2.35	1.78 2.30	1.75 2.22	1.72 2.18	1.69 2.13	1.67 2.09	1.65 2.06
29	4.18 7.60	3.33 5.42	2.93 4.54	2.70 4.04	2.54 3.73	2.43 3.50	2.35 3.33	2.28 3.20	2.22 3.08	2.18 3.00	2.14 2.92	2.10 2.87	2.05 2.77	2.00 2.68	1.94 2.57	1.90 2.49	1.85 2.41	1.80 2.32	1.77 2.27	1.73 2.19	1.71 2.15	1.68 2.10	1.65 2.06	1.64 2.03
30	4.17 7.56	3.32 5.39	2.92 4.51	2.69 4.02	2.53 3.70	2.42 3.47	2.34 3.30	2.27 3.17	2.21 3.06	2.16 2.98	2.12 2.90	2.09 2.84	2.04 2.74	1.99 2.66	1.93 2.55	1.89 2.47	1.84 2.38	1.79 2.29	1.76 2.24	1.72 2.16	1.69 2.13	1.66 2.07	1.64 2.03	1.62 2.01
32	4.15 7.50	3.30 5.34	2.90 4.46	2.67 3.97	2.51 3.66	2.40 3.42	2.32 3.25	2.25 3.12	2.19 3.01	2.14 2.94	2.10 2.86	2.07 2.80	2.02 2.70	1.97 2.62	1.91 2.51	1.86 2.42	1.82 2.34	1.76 2.25	1.74 2.20	1.69 2.12	1.67 2.08	1.64 2.02	1.61 1.98	1.59 1.96
34	4.13 7.44	3.28 5.29	2.88 4.42	2.65 3.93	2.49 3.61	2.38 3.38	2.30 3.21	2.23 3.08	2.17 2.97	2.12 2.89	2.08 2.82	2.05 2.76	2.00 2.66	1.95 2.58	1.89 2.47	1.84 2.38	1.80 2.30	1.74 2.21	1.71 2.15	1.67 2.08	1.64 2.04	1.61 1.98	1.59 1.94	1.57 1.91
36	4.11 7.39	3.26 5.25	2.86 4.38	2.63 3.89	2.48 3.58	2.36 3.35	2.28 3.18	2.21 3.04	2.15 2.94	2.10 2.86	2.06 2.78	2.03 2.72	1.98 2.62	1.93 2.54	1.87 2.43	1.82 2.35	1.78 2.26	1.72 2.17	1.69 2.12	1.65 2.04	1.62 2.00	1.59 1.94	1.56 1.90	1.55 1.87
38	4.10 7.35	3.25 5.21	2.85 4.34	2.62 3.86	2.46 3.54	2.35 3.32	2.26 3.15	2.19 3.02	2.14 2.91	2.09 2.82	2.05 2.75	2.02 2.69	1.96 2.59	1.92 2.51	1.85 2.40	1.80 2.32	1.76 2.22	1.71 2.14	1.67 2.08	1.63 2.00	1.60 1.97	1.57 1.90	1.54 1.86	1.53 1.84
40	4.08 7.31	3.23 5.18	2.84 4.31	2.61 3.83	2.45 3.51	2.34 3.29	2.25 3.12	2.18 2.99	2.12 2.88	2.07 2.80	2.04 2.73	2.00 2.66	1.95 2.56	1.90 2.49	1.84 2.37	1.79 2.29	1.74 2.20	1.69 2.11	1.66 2.05	1.61 1.97	1.59 1.94	1.55 1.88	1.53 1.84	1.51 1.81
42	4.07 7.27	3.22 5.15	2.83 4.29	2.59 3.80	2.44 3.49	2.32 3.26	2.24 3.10	2.17 2.96	2.11 2.86	2.06 2.77	2.02 2.70	1.99 2.64	1.94 2.54	1.89 2.46	1.82 2.35	1.78 2.26	1.73 2.17	1.68 2.08	1.64 2.02	1.60 1.94	1.57 1.91	1.54 1.85	1.51 1.80	1.49 1.78
44	4.06 7.24	3.21 5.12	2.82 4.26	2.58 3.78	2.43 3.46	2.31 3.24	2.23 3.07	2.16 2.94	2.10 2.84	2.05 2.75	2.01 2.68	1.98 2.62	1.92 2.52	1.88 2.44	1.81 2.32	1.76 2.24	1.72 2.15	1.66 2.06	1.63 2.00	1.58 1.92	1.56 1.88	1.52 1.82	1.50 1.78	1.48 1.75

Table 10 continued

Degrees of Freedom for Numerator

Degrees of Freedom for Denominator	1	2	3	4	5	6	7	8	9	10	11	12	14	16	20	24	30	40	50	75	100	200	500	∞
46	4.05 / 7.21	3.20 / 5.10	2.81 / 4.24	2.57 / 3.76	2.42 / 3.44	2.30 / 3.22	2.22 / 3.05	2.14 / 2.92	2.09 / 2.82	2.04 / 2.73	2.00 / 2.66	1.97 / 2.60	1.91 / 2.50	1.87 / 2.42	1.80 / 2.30	1.75 / 2.22	1.71 / 2.13	1.65 / 2.04	1.62 / 1.98	1.57 / 1.90	1.54 / 1.86	1.51 / 1.80	1.48 / 1.76	1.46 / 1.72
48	4.04 / 7.19	3.19 / 5.08	2.80 / 4.22	2.56 / 3.74	2.41 / 3.42	2.30 / 3.20	2.21 / 3.04	2.14 / 2.90	2.08 / 2.80	2.03 / 2.71	1.99 / 2.64	1.96 / 2.58	1.90 / 2.48	1.86 / 2.40	1.79 / 2.28	1.74 / 2.20	1.70 / 2.11	1.64 / 2.02	1.61 / 1.96	1.56 / 1.88	1.53 / 1.84	1.50 / 1.78	1.47 / 1.73	1.45 / 1.70
50	4.03 / 7.17	3.18 / 5.06	2.79 / 4.20	2.56 / 3.72	2.40 / 3.41	2.29 / 3.18	2.20 / 3.02	2.13 / 2.88	2.07 / 2.78	2.02 / 2.70	1.98 / 2.62	1.95 / 2.56	1.90 / 2.46	1.85 / 2.39	1.78 / 2.26	1.74 / 2.18	1.69 / 2.10	1.63 / 2.00	1.60 / 1.94	1.55 / 1.86	1.52 / 1.82	1.48 / 1.76	1.46 / 1.71	1.44 / 1.68
55	4.02 / 7.12	3.17 / 5.01	2.78 / 4.16	2.54 / 3.68	2.38 / 3.37	2.27 / 3.15	2.18 / 2.98	2.11 / 2.85	2.05 / 2.75	2.00 / 2.66	1.97 / 2.59	1.93 / 2.53	1.88 / 2.43	1.83 / 2.35	1.76 / 2.23	1.72 / 2.15	1.67 / 2.06	1.61 / 1.96	1.58 / 1.90	1.52 / 1.82	1.50 / 1.78	1.46 / 1.71	1.43 / 1.66	1.41 / 1.64
60	4.00 / 7.08	3.15 / 4.98	2.76 / 4.13	2.52 / 3.65	2.37 / 3.34	2.25 / 3.12	2.17 / 2.95	2.10 / 2.82	2.04 / 2.72	1.99 / 2.63	1.95 / 2.56	1.92 / 2.50	1.86 / 2.40	1.81 / 2.32	1.75 / 2.20	1.70 / 2.12	1.65 / 2.03	1.59 / 1.93	1.56 / 1.87	1.50 / 1.79	1.48 / 1.74	1.44 / 1.68	1.41 / 1.63	1.39 / 1.60
65	3.99 / 7.04	3.14 / 4.95	2.75 / 4.10	2.51 / 3.62	2.36 / 3.31	2.24 / 3.09	2.15 / 2.93	2.08 / 2.79	2.02 / 2.70	1.98 / 2.61	1.94 / 2.54	1.90 / 2.47	1.85 / 2.37	1.80 / 2.30	1.73 / 2.18	1.68 / 2.09	1.63 / 2.00	1.57 / 1.90	1.54 / 1.84	1.49 / 1.76	1.46 / 1.71	1.42 / 1.64	1.39 / 1.60	1.37 / 1.56
70	3.98 / 7.01	3.13 / 4.92	2.74 / 4.08	2.50 / 3.60	2.35 / 3.29	2.23 / 3.07	2.14 / 2.91	2.07 / 2.77	2.01 / 2.67	1.97 / 2.59	1.93 / 2.51	1.89 / 2.45	1.84 / 2.35	1.79 / 2.28	1.72 / 2.15	1.67 / 2.07	1.62 / 1.98	1.56 / 1.88	1.53 / 1.82	1.47 / 1.74	1.45 / 1.69	1.40 / 1.63	1.37 / 1.56	1.35 / 1.53
80	3.96 / 6.96	3.11 / 4.88	2.72 / 4.04	2.48 / 3.56	2.33 / 3.25	2.21 / 3.04	2.12 / 2.87	2.05 / 2.74	1.99 / 2.64	1.95 / 2.55	1.91 / 2.48	1.88 / 2.41	1.82 / 2.32	1.77 / 2.24	1.70 / 2.11	1.65 / 2.03	1.60 / 1.94	1.54 / 1.84	1.51 / 1.78	1.45 / 1.70	1.42 / 1.65	1.38 / 1.57	1.35 / 1.52	1.32 / 1.49
100	3.94 / 6.90	3.09 / 4.82	2.70 / 3.98	2.46 / 3.51	2.30 / 3.20	2.19 / 2.99	2.10 / 2.82	2.03 / 2.69	1.97 / 2.59	1.92 / 2.51	1.88 / 2.43	1.85 / 2.36	1.79 / 2.26	1.75 / 2.19	1.68 / 2.06	1.63 / 1.98	1.57 / 1.89	1.51 / 1.79	1.48 / 1.73	1.42 / 1.64	1.39 / 1.59	1.34 / 1.51	1.30 / 1.46	1.28 / 1.43
125	3.92 / 6.84	3.07 / 4.78	2.68 / 3.94	2.44 / 3.47	2.29 / 3.17	2.17 / 2.95	2.08 / 2.79	2.01 / 2.65	1.95 / 2.56	1.90 / 2.47	1.86 / 2.40	1.83 / 2.33	1.77 / 2.23	1.72 / 2.15	1.65 / 2.03	1.60 / 1.94	1.55 / 1.85	1.49 / 1.75	1.45 / 1.68	1.39 / 1.59	1.36 / 1.54	1.31 / 1.46	1.27 / 1.40	1.25 / 1.37
150	3.91 / 6.81	3.06 / 4.75	2.67 / 3.91	2.43 / 3.44	2.27 / 3.13	2.16 / 2.92	2.07 / 2.76	2.00 / 2.62	1.94 / 2.53	1.89 / 2.44	1.85 / 2.37	1.82 / 2.30	1.76 / 2.20	1.71 / 2.12	1.64 / 2.00	1.59 / 1.91	1.54 / 1.83	1.47 / 1.72	1.44 / 1.66	1.37 / 1.56	1.34 / 1.51	1.29 / 1.43	1.25 / 1.37	1.22 / 1.33
200	3.89 / 6.76	3.04 / 4.71	2.65 / 3.88	2.41 / 3.41	2.26 / 3.11	2.14 / 2.90	2.05 / 2.73	1.98 / 2.60	1.92 / 2.50	1.87 / 2.41	1.83 / 2.34	1.80 / 2.28	1.74 / 2.17	1.69 / 2.09	1.62 / 1.97	1.57 / 1.88	1.52 / 1.79	1.45 / 1.69	1.42 / 1.62	1.35 / 1.53	1.32 / 1.48	1.26 / 1.39	1.22 / 1.33	1.19 / 1.28
400	3.86 / 6.70	3.02 / 4.66	2.62 / 3.83	2.39 / 3.36	2.23 / 3.06	2.12 / 2.85	2.03 / 2.69	1.96 / 2.55	1.90 / 2.46	1.85 / 2.37	1.81 / 2.29	1.78 / 2.23	1.72 / 2.12	1.67 / 2.04	1.60 / 1.92	1.54 / 1.84	1.49 / 1.74	1.42 / 1.64	1.38 / 1.57	1.32 / 1.47	1.28 / 1.42	1.22 / 1.32	1.16 / 1.24	1.13 / 1.19
1000	3.85 / 6.66	3.00 / 4.62	2.61 / 3.80	2.38 / 3.34	2.22 / 3.04	2.10 / 2.82	2.02 / 2.66	1.95 / 2.53	1.89 / 2.43	1.84 / 2.34	1.80 / 2.26	1.76 / 2.20	1.70 / 2.09	1.65 / 2.01	1.58 / 1.89	1.53 / 1.81	1.47 / 1.71	1.41 / 1.61	1.36 / 1.54	1.30 / 1.44	1.26 / 1.38	1.19 / 1.28	1.13 / 1.19	1.08 / 1.11
∞	3.84 / 6.64	2.99 / 4.60	2.60 / 3.78	2.37 / 3.32	2.21 / 3.02	2.09 / 2.80	2.01 / 2.64	1.94 / 2.51	1.88 / 2.41	1.83 / 2.32	1.79 / 2.24	1.75 / 2.18	1.69 / 2.07	1.64 / 1.99	1.57 / 1.87	1.52 / 1.79	1.46 / 1.69	1.40 / 1.59	1.35 / 1.52	1.28 / 1.41	1.24 / 1.36	1.17 / 1.25	1.11 / 1.15	1.00 / 1.00

*From Biometrika Tables for Statisticians, Vol. I, by permission of the Biometrika Trustees.

Table 11 Critical Values for Spearman Rank Correlation, r_s

For a right- (left-) tailed test, use the positive (negative) critical value found in the table under significance level for a one-tailed test. For a two-tailed test, use both the positive and negative of the critical value found in the table under significance level for a two-tailed test, n = number of pairs.

	Significance level for a one-tailed test at			
	0.05	0.025	0.005	0.001
	Significance level for a two-tailed test at			
n	0.10	0.05	0.01	0.002
5	0.900	1.000		
6	0.829	0.886	1.000	
7	0.715	0.786	0.929	1.000
8	0.620	0.715	0.881	0.953
9	0.600	0.700	0.834	0.917
10	0.564	0.649	0.794	0.879
11	0.537	0.619	0.764	0.855
12	0.504	0.588	0.735	0.826
13	0.484	0.561	0.704	0.797
14	0.464	0.539	0.680	0.772
15	0.447	0.522	0.658	0.750
16	0.430	0.503	0.636	0.730
17	0.415	0.488	0.618	0.711
18	0.402	0.474	0.600	0.693
19	0.392	0.460	0.585	0.676
20	0.381	0.447	0.570	0.661
21	0.371	0.437	0.556	0.647
22	0.361	0.426	0.544	0.633
23	0.353	0.417	0.532	0.620
24	0.345	0.407	0.521	0.608
25	0.337	0.399	0.511	0.597
26	0.331	0.391	0.501	0.587
27	0.325	0.383	0.493	0.577
28	0.319	0.376	0.484	0.567
29	0.312	0.369	0.475	0.558
30	0.307	0.363	0.467	0.549

From G. J. Glasser and R. F. Winter, "Critical Values of the Coefficient of Rank Correlation for Testing the Hypothesis of Independence," *Biometrika, 48,* 444 (1961). Printed by permission of Biometrika Trustees.

Other Useful Tables

Table 8-2 Some Levels of Confidence and Their Corresponding Critical Values

Level of Confidence c	Critical Value z_c
0.75	1.15
0.80	1.28
0.85	1.44
0.90	1.645
0.95	1.96
0.99	2.58

Table for Hypothesis Testing   Critical Values of Z in Hypothesis Testing

Type of Test	$\alpha = 0.05$	$\alpha = 0.01$
Left-ailed Test	$z_0 = -1.645$	$z_0 = -2.33$
Right-Tailed Test	$z_0 = 1.645$	$z_0 = 2.33$
Two-Tailed Test	$\pm z_0 = \pm 1.96$	$\pm z_0 = \pm 2.58$